THE
MECHANIC'S COMPANION

By
Peter Nicholson
1850

A Facsimile Reprint
From Our Personal Library To Yours

Introduction by Gary Roberts

The Toolemera Press

www.toolemera.com

The Mechanic's Companion
by Peter Nicholson
1850

International Standard Book Number
ISBN: 978-0-9831500-0-8
(Trade Paperback)

Published by
The Toolemera Press
Dedham, Massachusetts
U.S.A. 02026

http://toolemera.com

Manufactured in the United States of America

Introduction
by Gary Roberts

The Mechanic's Companion was first published in 1812 as *Mechanical Exercises* and then republished in 1831 under the present title. Peter Nicholson's compendium of technical advise has maintained it's position as a primary reference source through to the present day.

The Toolemera Press reprint is taken from an 1850 edition. In addition to the 1850 copy, we have 1832 and 1849 editions. Comparing these three books revealed that the 1849 and 1850 editions had been printed from the same stereotype plates as were used in the 1832 edition.

1812 Book Review

The Literary Panorama: Being A Compendium Of National Papers and Parliamentary Reports, Illustrative Of The History, Statistics, and Commerce of the Empire: A Universal Epitome of Interesting and Amusing Intelligence from All Quarters of the Globe; A Review Of Books, and Magazines of Varieties, Forming An Annual Register.
Vol. XL. J. Taylor. London. 1812

Mechanical Exercises; or, the Elements of Practice of Carpentry, Joinery, Bricklaying, Masonry, Slating, Plastering, Painting, Smithing and Turning. Containing a full Description of the Tools belonging to each Branch of Business; and copious Directions for the Use. With an Explanation of the Terms used in each Art; and an Introduction by Thirty-nine Copper Plates.
By Peter Nicholson. 8vo. Pp. 420. Price 18s.
J. Taylor. London. 1812

"More than a century has elapsed since Moxon's ingenious work on the building art, called "Mechanical Exercises" was first published. It long continued useful and popular. The present state of practice allowed very well of a similar work with improvements. Mr. Nicholson has followed the steps of his predecessor, and treats separately of the arts enumerated in his title page. The book is principally intended for the service of young men coming from country to town, who will find many operations here, with which they have not been familiar. The readiest way of performing these, must indeed, be learned from practice; but a work like the present may tend advantageously to lessen that ignorance of which they are sensible, and to place them more on a level with their fellow workmen. The arts concerned in building are sufficiently connected with one another to justify a wish for acquaintance with more than one; and when directions are given, it is no detriment to be able to give them in language understood by the workman. We doubt whether Mr. N. has fully executed his intentions when he proposed to compile lists of the terms used in each art. From a practical man we should have expected a vocabulary of which a future Johnson might avail himself; - it is to be expected from a practical man only; for neither Greek nor Latin can assist in this matter. Many a good classic scholar does not the difference between "carpentry," and "joinery," but let him contract for a house, expecting to find all the wood work completed and finished, though he bargains only for the "carpentry," he will soon find his knowledge improved by experience. Mr. N. should have had still further compassion on the ignorant. He might have introduced his treatise on carpentry, by a few words on the nature and quality of woods: he notices only oak and fir; why not elm, beech, &c. with others used in turnery, of which art he also treats. Attention to the properties of things, whether materials, or implements, carried throughout his volume, would have materially improved it: a knowledge of the goodness of tools, is of the first importance to a workman; and is undoubtedly one great cause of the excellence unanimously attributed by foreigners to productions of British skill and industry.

The plates of this work are useful: they describe tools and operations, which *some* acquaintance with practice will render beneficial.

The difficulty of describing the *simplest* of things, in words, was the apology of Dr. Johnson for his famous definition of "*Network*" (Any thing reticulated or decussated at equal distances with interstices between intersections.) Let our readers try by way of amusement, to define any implement to which they are accustomed; and they will feel the force of the doctor's vindication of himself: - what precise ideas will they convey *by words*? - The following is Mr. N.'s definition of the well known tool a *Gimblet* - If this writer employs so many words to render a thing intelligible, which, after all, unless we know it perfectly beforehand we should not understand, where there is the wonder that a lexicographer should be at a loss, a writer who is so much better acquainted with words than it is possible he should be with things?

The Gimblet is a piece of steel of a cylindric form, having a transverse handle at the upper end, and at the other, a worm or screw; and a cylindric cavity called the cup above the screw; forming in its transverse section, a crescent. Its use is to bore small holes; the screw draws it forward in the wood, in the act of boring, while it is turned round by the handle; the angle formed by the exterior and interior cylinders, cuts the fibres across, and the cup contains the core of wood so cut: the gimblet is turned round by application of the fingers, on alternate sides of the wooden lever at the top.

Those clauses of the building act which relate to each trade are separately transcribed, and placed after each division. This may preserve many workmen serious evils, who inadvertently might undertake private jobs."

Biography

Peter Nicholson, 1765-1844

Born in Scotland and having lived in both Scotland and England, Nicholson was exposed to the emerging architectural influence of Grecian sensibilities over the prevailing Gothic taste. At an early age his proficiency in mathematics and skills in the drawing of local buildings indicated the direction his professional life would eventually take. Initially entering into training with his father, a stone mason, he found his interests to lie in the working of wood and so apprenticed to a cabinet-maker.

Upon completion of his four year apprenticeship, he worked as a journeyman in Edinburgh, Scotland, and London, England. While pursuing his journeyman tasks, Nicholson secured a position as a teacher at an evening school. He did so well in this new profession that he left cabinet-work and began his new career of builder (architect) and author.

At the age of 27, Nicholson authored and engraved the plates for "The Carpenter's New Guide". He went on to write "The Student's Instructor", "The Carpenter and Joiner's Assistant", and 24 more titles on architecture and related subjects. During his lifetime, his books were essential to the libraries of beginning as well as experienced builders and architects.

Nicholson's books were predominantly intended for use by the practical carpenter and builder more so than by the theoretical architect. His hands-on experience in the trades as well as his unique aptitude for mathematics prepared him for his career as a self-taught architect, builder and author.

Toolemera Press Facsimile Reprints

The Toolemera Press reprints classic books and ephemera on early tools, trades and industries. We will only reprint items held in our personal library. We will never use a source document from any online document depository. The Toolemera Press manages every aspect of the publishing process. All imaging is accomplished either in-house or by contract with respected document imaging services. We use Print-On-Demand to keep pricing affordable.

http://toolemera.com

contact@toolemera.com

NICHOLSON'S

MECHANIC'S COMPANION.

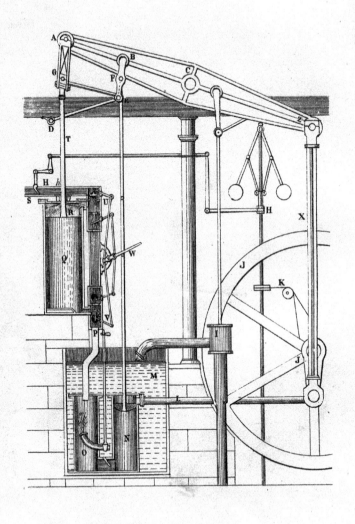

WATT'S DOUBLE ACTING ENGINE. *(See page 339)*

THE

MECHANIC'S COMPANION;

OR, THE

ELEMENTS AND PRACTICE

OF

CARPENTRY, JOINERY, BRICKLAYING, MASONRY,
SLATING, PLASTERING, PAINTING,
SMITHING, AND TURNING,

COMPREHENDING THE LATEST IMPROVEMENTS;

AND CONTAINING A FULL

DESCRIPTION OF THE TOOLS

BELONGING TO EACH BRANCH OF BUSINESS, WITH COPIOUS DIRECTIONS
FOR THEIR USE;

AN EXPLANATION OF THE TERMS USED IN EACH ART;

AND AN

INTRODUCTION TO PRACTICAL GEOMETRY.

———————

BY PETER NICHOLSON.

———————

TO WHICH IS ADDED

AN ESSAY ON THE STEAM-ENGINE,

ITS MANAGEMENT, USES, &c.

WITH FORTY-SIX ENGRAVINGS.

PHILADELPHIA:
PUBLISHED BY F. BELL.
1850.

PREFACE.

MORE than a century has elapsed since an in-genious and useful work on the Arts connected with BUILDING was published under the title of Mechanical Exercises, by the celebrated JOSEPH MOXON: that it was both useful and popular the various editions testify, and at this time it is become scarce and rarely to be met with. It can be no disparagement to its ingenious author, to say that the progress of science, and the changes in matters of art have rendered the work obsolete and useless. It treated on Smithing, Joinery, Carpentry, Turning, Bricklaying, and Dialling.

I have followed the excellent plan of Moxon and treated each art distinctly: I have first described the several tools belonging to each branch of business, next the methods of performing the various manual

operations or exercises, to which they are applicable,
these are further illustrated and explained by numerous plates: the descriptions are made as plain and
familiar as possible; and there are few operations
but will be found fully and clearly explained: finally
to each is added an Index and extensive Glossary of
terms used by workmen in each art, with references
also to the plates: and it has been my endeavour
that the description with its definition should be
clear, and show the connection between the science
and the art, thereby producing a pleasing and lasting
effect upon the mind.

The arts treated of are as follow: CARPENTRY,
JOINERY, BRICKLAYING, MASONRY, SLATING, PLASTER-
ING, PAINTING, SMITHING, AND TURNING, the whole
preceded by a slight introduction to PRACTICAL
GEOMETRY, and illustrated by forty copper-plates.

These exercises commence with those arts which
work in wood, namely, Carpentry and Joinery which
are much alike in their tools and modes of working :
then comes Bricklaying, which with Carpentry are
certainly the most essential of all in the construction
of a building.

Masonry and Bricklaying are in reality branches of the same art, and both founded upon principles truly Geometrical, yet I have given the precedence to Bricklaying, because it is of the most general use in this country; yet it is generally admitted, that Masonry is the more dignified art of the two, or indeed of all the arts concerned in the formation of an edifice. On that difficult and intricate subject, the Theory of Arches, I have endeavoured to give a familiar, and I hope a satisfactory illustration.

Slating comes next to cover in the building: then Plastering, which is used in the finishing of buildings, and furnishes the interior with elegant decorations, and conduces both to the health and comfort of the inhabitants: Painting is not less useful than ornamental; it adds to the elegance of buildings, and tends to the preservation of the materials, whether wood or plaster.

Smithing or Smithry is extensively useful in almost every department of art as well as building; by it are made the tools which perform all the operations of the before mentioned arts, and therefore, though last, should not be least in our esteem. The use of

iron has also of late years been very much extended;
in wheels for machinery, Iron Bridges, Rail roads,
Boats, Roofs, Floors, and various other articles not
necessary to enumerate here.

Turning is a curious Mechanical Exercise, and
though not absolutely necessary in building, may be
employed with advantage in many of its decorations.
In this article I have given a legitimate definition of
elliptic turning, by which, its principles are deduced
to be that of the ellipsegraph or common trammel,
and this without entering into further demonstration.
This art is illustrated by plates, showing the principles of the machines, as well as by views of the
machines and tools

As the practice of the arts here treated of, is
founded in Geometry, and as the descriptions of
the materials and of the tools may be referred to the
several figures of that science, I have prefixed to
the work such definitions as are necessary to the
comprehension of any drawing or design, which is
to be executed, accompanied by many useful problems, which will enable the mechanic to understand
the configuration of its several parts in practice,

and to perform many useful problems upon true scientific principles. The problems for setting out work upon the ground, and those for reducing drawings to any scale or proportion, even without knowing the scale of the original drawing, will be found interesting, and very useful in practice.

This work, which treats of the first rudiments of practice, will be found particularly interesting and useful to gentlemen who practise, or are fond of the mechanical exercises, and to young men or apprentices in any of the professions, though, on some occasions, the older workmen may be benefitted by a perusal. The terms introduced are those in general use amongst workmen in London: and on this account it will be of essential service to young men coming to the metropolis. An art cannot be taught but by its proper terms. Other branches of art might have been introduced into this work, but those here treated of are intimately connected with each other, and have a natural affinity, and will, it is presumed, form upon the whole, a very interesting work to young mechanics; those who wish for further information in the building art, and particularly

B

on what relates to Geometrical Construction, may consult my other publications on Practical Carpentry.

Every art is improved by the emulation of its competitors: it is therefore the ardent hope of the author that the reader may not be disappointed of meeting with abundance of that information which his mind may be desirous to obtain.

PETER NICHOLSON.

PRACTICAL GEOMETRY.

GEOMETRY is the science of extension and magnitude: by Geometry the various angles of a building and the position of its sides are determined, as a square, a cube, a triangle, &c.: Boards and all Tools used by the Carpenter and Joiner are geometrical constructions: by Geometry all kinds of roofs and various other things laying in oblique angles are determined: the proper construction of all sòrts of arches and groins depend entirely upon the principles of Geometry. I have, therefore, prefaced this work with an explanation and definition of such geometrical figures as will frequently occur in carrying on of works, and which are therefore necessary to be well known by all artizans and workmen, as well as by those who may superintend them: this slight introduction to Geometry will also be useful to all persons who wish to understand the practice and descriptions of the handy-works herein explained.

Geometry is the science of extension, and magnitude, and consists of theory and practice.

The theoretical part is founded upon the reasoning of self-evident principles; it demonstrates the construction, and shows the properties of regularly defined figures. The theory is the foundation of the practical part; and without a knowledge of it, no invention to any degree certain can be made. The use of Geometry is not confined only to speculative truths in Mathematics, but the operations of mechanical arts owe their perfection to it ; drawing and setting out every description of work, are entirely dependent upon it.

DEFINITIONS.

1. A point is that which has position, but not magnitude.

2. A line is the trace of a point, or that which would be described by the progressive motion of a point, and consequently has length only.

3. A superfices has length and breadth.

4. A solid is a figure of three dimensions, having length, breadth, and thickness. Hence surfaces are extremities of solids, and lines the extremities of surfaces, and points the extremities of lines.

If two lines will always coincide, however applied, when any two points in the one coincides with the two points in the other, the two lines are called straight lines, or otherwise right lines.

A curve continually changes its direction between its extreme points, or has no part straight.

Parallel lines are always at the same distance, and will never meet, though ever so far produced. Oblique right lines change their distance, and would meet if produced.

One line is perpendicular to another, when it inclines no more to one side than another.

A straight line is a tangent to a circle, when it touches the circle without cutting when both are produced.

An angle is the inclination of two lines towards one another in the same plane, meeting in a point.

Angles are either right, acute, or oblique.

A right angle is that which is made by one line perpendicular to another, or when the angles on each side are equal.

An acute angle is less than a right angle.

An obtuse angle is greater than a right angle.

A plane is a surface with which a straight line will every where coincide : and is otherwise called a straight surface.

Plane figures, bounded by right lines, have names according to the number of their sides, or of their angles, for they have as many sides as angles : the least number is three.

An equilateral triangle is that whose three sides are equal.

An isosceles triangle has only two sides unequal.

A scalene triangle has all sides unequal.

A right-angle triangle has only one right angle.

Other triangles are oblique-angled, and are either obtuse or acute.

An acute-angled triangle has all its angles acute.

An obtuse-angled triangle has one obtuse angle.

A figure of four sides, or angles, is called a quadrilateral, or quadrangle.

A parallelogram is a quadrilateral, which has both pairs of its opposite sides parallel, and takes the following particular names:

A rectangle is a parallelogram, having all its angles right ones.

A square is an equilateral rectangle, having all its sides equal, and all its angles right ones.

A rhombus is an equilateral parallelogram whose angles are oblique.

A rhomboid is an oblique-angled parallelogram, and its opposite sides only are equal.

A trapezium is a quadrilateral, which has neither pair of its sides parallel.

A trapezoid hath only one pair of its opposite sides parallel.

Plane figures having more than four sides, are in general called polygons, and receive other particular names according to the number of their sides or angles.

A pentagon is a polygon of five sides, a hexagon of six sides, a heptagon seven, an octagon eight, an eneagon nine, a decagon ten, an undecagon eleven, and a dodecagon twelve sides.

A regular polygon has all its sides, and its angles equal ; and if they are not equal, the polygon is irregular.

An equilateral triangle is also a regular figure of three sides, and a square is one of the four; the former being called a trigon, and the latter a tetragon.

A circle is a plane figure, bounded by a curve line, called the circumference, which is every where equi-distant, from a certain point within, called its centre.

The radius of a circle is a right line drawn from the centre to the circumference.

A diameter of a circle is a right line, drawn through the centre terminating on both sides of the circumference

An arc of a circle is any part of the circumference.

A chord is a right line joining the extremities of an arc.

A segment is any part of a circle bounded by an arc and its chord.

A semicircle is half a circle, or a segment cut off by the diameter.

A sector, is any part of a circle bounded by an arc, and two radii, drawn to its extremities.

A quadrant, or quarter of a circle, is a sector having a quarter part of the circumference for its arc, and the two radii perpendicular to each other.

The height or altitude of any figure is a perpendicular, let fall from an angle or its vertex, to the opposite side, called the base.

The measure of any right lined angle, is an arc of any circle contained between the two lines which form the angle, the angular point being the centre.

A solid is said to be cut by a plane, when it is divided into two parts, of which the common surface of separation is a plane, and this plane is called a section.

DEFINITIONS OF SOLIDS.

A prism is a solid, the ends of which are similar, and equal parallel planes, and the sides parallelograms.

If the ends of the prism are perpendicular to the sides, the prism is called a right prism.

If the ends of the prism are oblique to the sides, the prism is called an oblique prism.

If the ends and sides are equal squares, the prism is called a cube,

If the base or ends are parallelograms, the solid is called a parallelopiped.

If the bases and sides are rectangles, the prism is called a rectangular prism.

If the ends are circles, the prism is called a cylinder.

If the ends or bases are ellipsis, the prism is called a cylindroid.

A solid, standing upon any plane figure for its base, the sides of which are plane triangles, meeting in one point, is called a pyramid.

The solid is denominated from its base, as a triangular pyramid is one upon a triangular base, a square pyramid one upon a square base, &c.

If the base is a circle or an ellipsis, then the pyramid is called a cone.

If a solid be terminated by two dissimilar parallel planes as ends, and the remaining surfaces joining the ends be also planes, the solid is called a prismoid.

If a part of a pyramid next to the vertex be cut off by a plane parallel to the base, the portion of the pyramid contained between the cutting plane and the base is called the frustum of a pyramid.

A solid, the base of which is a rectangle, the four sides joining the base plane surfaces, and two opposite ones meet in a line parallel to the base, is called a cuneus or wedge.

A solid terminated by a surface which is every where equally distant, from a certain point within it, is called a sphere or globe.

If a sphere be cut by any two planes, the portion contained between the planes is called a zone, and each of the parts contained by a plane and the curved surface is called a segment.

If a semi-ellipsis, having an axis for its diameter, be revolved round this axis until it come to the place whence the motion began, the solid formed by the circumvolution is called a spheroid.

If the spheroid be generated round the greater axis, the solid is called an oblong spheroid.

If the solid be generated round the lesser axis, the solid is called an oblate spheroid.

A solid of any of the above structures, hollow within, so as to contain a solid of the same structure, is called a hollow solid.

PLATE I.

A an acute angle.

B two lines inclined, and would meet and form an angle if produced.

C a perpendicular *c d* is said to be perpendicular to *a b*, and the angles *c d a, c b d* are both right angles.

D several angles meeting at a point; when this is the case, each is denoted by three letters, the right angle is the criterion of judging of every other angle; *d b c* is a right angle; *a b c* an obtuse angle, *e b c* an acute angle.

E a right angle.

F an acute angle, being less than a right angle

G an obtuse angle, being greater than a right angle.

H, I, K, L triangles.

H an equilateral triangle all the three sides *a b, b c, c a* being equal.

I an isosceles triangle, *a b* and *b c* being only equal.

K a scalene triangle, all the sides being unequal.

L a right angled triangle.

M, N, O, P, Q, R quadrilaterals or quadrangles, M N O P are parallelograms; M N rectangles; M an oblong; N a square; O a rhomboid; P a rhombus; Q a trapezium; and R a trapezoid.

T, U, V polygons, T a pentagon, U a hexagon, and V an octagon.

W a circle, *a* the centre, *b* a point in the circumference, *a b* a radius.

X a circle, *c* the centre, *d* and *e* points in the circumference, *d e a* a diameter, or a chord passing through the centre.

Geometry Plate I.

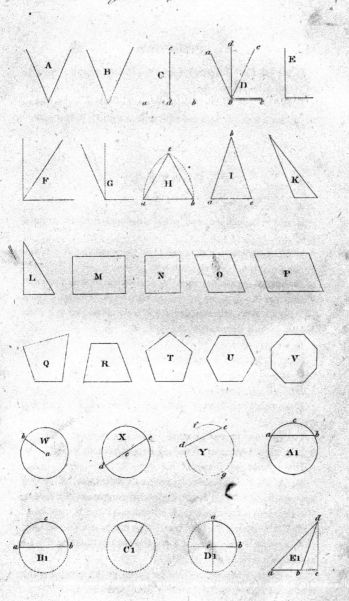

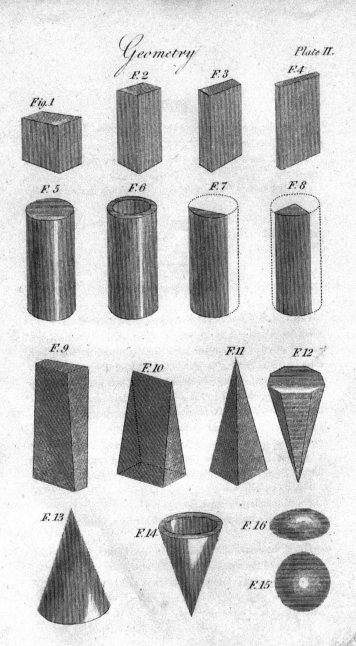

Geometry

Plate II.

Fig.1   F.2   F.3   F.4

F.5   F.6   F.7   F.8

F.9   F.10   F.11   F.12

F.13   F.14   F.16

F.15

Y a circle, *d* and *e* points in the circumference, *d e* a chord ; *d f e* the less segment, and *d g e* the greater.

A 1, B 1 segments, *a c b*, *a &b* arcs, *a b*, *a.b* chords; B 1 a semicircle.

C 1, D 1 sectors, D 1 a quadrant, *c a*, *c b* radii at right angles, *a b* arc.

E 1 a triangle, *a b*, *b d*, *d a* the sides, *a b* the base, *d c a* perpendicular to the base called the altitude.

PLATE II.

Fig. 1, 2, 3, 4 are all parallelopipeds and consist of six sides ; when two opposite sides are perpendicular to the other four, the parallelopiped is denominated a rectangular prism, and if the four sides be equal rectangles, the prism is called a square prism as fig. 1, 2; and if all the four sides are equal squares, the prism is called a cube, as fig. 1. The reason why called a parallelopiped is because each pair of opposite sides are parallel planes. The structure of a rectangular prism occurs more frequently in the practice of carpentry and joinery than any other form whatever, all timbers and boards for the use of building are cut into this form. Doors, shutters, &c. are thin rectangular prisms, as fig. 4.

Fig. 5 is a cylinder.

Fig. 6 a hollow cylinder.

Fig. 7 the section of a cylinder cut off by a plane parallel to the axis.

Fig. 8 the sector of a cylinder contained by two planes forming an angle, and the curved surface of the cylinder; the line of concourse of the planes being parallel to the axis of the cylinder.

Fig. 9 a prismoid; the ends of chisels which contain the cutting part is of this form.

Fig. 10 a wedge; the end of a chisel contained by the face and the basil are of this form.

c

Fig. 11 a square pyramid.

Fig. 12 an octagonal pyramid inverted

Fig. 13 a cone.

Fig. 14 inverted hollow cone.

Fig. 15 a sphere.

Fig. 16 a spheroid.

———

PROB. I. *From a given point in a given straight line, to erect a perpendicular.* PL. 3. FIG. I.

Let F F be the given straight line, and C the given point. Take any two equal distances C *a* and C *b* on each side of the point C : from the points *a* and *b* with any equal radii greater than C *a* or C *b*, describe arcs cutting each other in D. Draw D C and it will be the perpendicular required.

———

PROB. II. *To let fall a perpendicular from a given point to a given straight line.* PL. 3. FIG. 2.

Let C be the given point and E F the given straight line. From the point C describe an arc cutting E F at *a* and *b*. With any equal radii greater than the half of *a b* describe arcs cutting each other at D. Draw C D and it will be the perpendicular required.

———

PROB. III. *When the point is at or near the end of the line. Method first,* PL. 3. FIG. 3.

Let C be the given point, E F the given line. In E F take any point *a* and with the radius *a* C describe an arc C D. Take any other point *b* in E F, and with the distance *b* C describe an arc, cutting the arc C D, at C and D, draw C D and it is the perpendicular required .

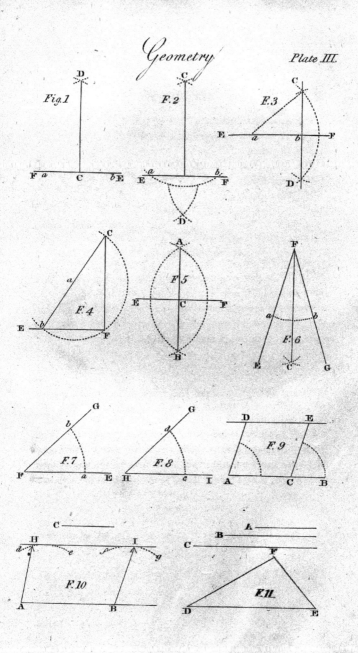

Fig.1 F.2 F.3 F.4 F.5 F.6 F.7 F.8 F.9 F.10 F.11

Prob. iv. *To draw a perpendicular from a point at the end of a line.* Pl. 3. Fig. 4.

Let E F be the given straight line, and F the given point. Take any point *a* above the line and with the radius *a* C describe an arc C F *b* cutting E F at *b*. Draw *b a* C; then draw C F and it will be the perpendicular required.

Prob. v. *To bisect a straight line.* Pl. 3. Fig. 5.

Let E F be the given straight line. From E and F as centres, and with any distance greater than the half of E F as radii, describe two arcs cutting each other at A and B. Draw A B cutting E F at C, then E F is bisected in C.

Prob. vi. *To bisect a given angle.* Pl. 3. Fig. 6.

Let E F G be the given angle. From the point F describe an arc *a b* cutting F E and F G at the points *a* and *b*, also from the points *a* and *b*, with the same radius, or any other equal radii, describe arcs cutting each other in C. Draw F C and it will be-sect the angle as required. That is, the angle E F G is divided into two equal angles E F C and C F G.

Pron. vii. *To make an angle equal to a given angle.* Pl. 3. Fig. 7 and 8.

Let E F G be the given angle. Draw the straight line H I. From the point F describe an arc *a b* cutting E F and F G at the points *a* and *b*. From H as a centre, with the same radius, describe an arc *c d* cutting H I at *c*. Make *c d* equal to *a b*. Draw H *d* G and the angle I H G is equal to E F G, as required.

PROB. VIII. *Through a given point to draw a line parallel to a given right line.* PL. 3. FIG. 9.

Let A B be the given right line, and D the given point. Draw any right line D A ; in A B take any point *c* and make the angle B *c* E equal to the angle B A D, make *c* E equal to A D ; draw Đ E, then D E is parallel to A B.

PROB. IX. *To draw a line parallel to another line at a given distance.* PL. 3. FIG. 10.

Let A B be the given right line, C the given distance from any two points in A B, as A and B as centres describe two arcs *d* H *e* and *f* I *g*. Draw H I to touch the arcs at the points H and I ; and H I is parallel to A B and at a given distance C.

PROB. X. *Three straight lines, of which any two are greater than the third, being given to describe a triangle, the sides of which will be respectively equal to the three given lines.* PL. 3. FIG. 11.

Let the three straight lines be A B C : Make D E equal to C, from D as a centre with the distance of B describe an arc at F. From E as a centre with the distance A describe another arc, cutting the former at F. Join F D and F E ; and D E F is the triangle required.

PROB. XI. *The side of an equilateral triangle being given to describe the triangle.* PL. 4. FIG. 1.

Let A be the given side. Place A upon any straight line B C and with the same extent from the points B and C as centres

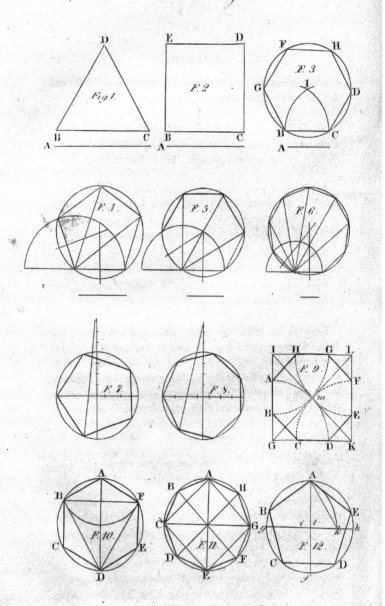

describe arcs, cutting each other in D. Join D B and D C, and B C D is the equilateral triangle required.

PROB. XII. *To describe a square, the sides of which shall be equal to a given right line.* PL. 4. FIG. 2.

Let A be the given right line, which place upon any straight line B C. Make the angle C B E a right angle, and B E equal to B C through the points E and C. Draw E D and D C parallel to B C, and B E, and B C D E is the square required.

PROB. XIII. *To describe a hexagon, the sides of which shall be equal to a given line.* PL. 4. FIG. 3.

Let A be the given line, which place upon any straight line B C. From the points B and C, with the distance B C describe arcs cutting each other at I. With the distance I B or I C describe the circle B C D E F G, then apply the side B C successively to the circumference as chords, the circumference will be divided into equal parts, and the hexagon formed as required.

PROB. XIV. *To describe any regular polygon, the sides of which shall be equal to a given line.* PL. 4. FIG. 4.

Set the given line upon any other convenient line, and with a radius equal to the given line, describe a semicircle upon this line. Divide the semicircle into as many equal parts as are to be sides in the polygon; then the half of the diameter is one side of the polygon, through the centre of the semicircle, and through the second division from the other end of the diameter draw another right line, which will form an adjoining side to the former; bisect each of these adjoining sides by perpendiculars, and the meeting of these perpendiculars will give the centre of a circle, which will contain the straight line given.

Fig. 4 is an example of a pentagon.

Fig. 5 is an example of a hexagon.

Fig. 6 is an example of an eneagon.

PROB. XV. *To inscribe a polygon in a given circle.* PL. 4. FIG. 7, 8

Draw the diameter of the circle, and another diameter at right angles, produce this last diameter so that the part produced shall be three quarters of the radius; divide the first diameter into as many equal parts as the polygon is to consist of sides: through the second division, and the extremity of the part produced of the other diameter, draw a line to cut the circumference without the points, the chord of the arc intercepted between the point in the circumference thus found and the diameter, applied successively to the arc, as other chords will form the polygon required.

Fig. 7 example in a pentagon, Fig. 8 example in an octagon.

PROB. XVI. *A square being given to form an octagon, of which four of the sides at right angles to each other, shall be common to the middle parts of the sides of the square.* PL. 4. FIG. 9.

Let I G K L be the square given. Draw the diagonals I K and G L cutting each other at *m;* from the centres I, G, K, L and with the radius I *m,* or G *m,* &c. describe arcs G *m* B, A *m* D, C *m* F, E *m* H cutting the sides of the square, at A, B, C, D, E, F, G, H; join B C, D E, F G, H A and A B C D E F G H will be the polygon as required.

PROB. XVII. *In a given circle to inscribe a hexagon or an equaliteral.*
PL. 4. FIG. 10.

Apply the radius successively as chords A B, B C, C D,
D E, E F, F A, and A B C D E F A will be the hexagon.
From A with the radius A B or A F describe the arc B F,
Join the chord B F. Make B D equal to B F; and join D F
and B F D is the equilateral triangle required.

———

PROB. XVIII. *In a given circle to inscribe a square or an octagon.*
PL. 4. FIG. 11.

Let A B C D E F G H A be the circle. Draw the diameters
A E and C G at right angles. Join A C, C E, E G, G A and
A C E G A will be the square required.

Bisect any two adjacent angles by diameters, and the whole
circumference will be divided into eight equal parts, A B, B C,
C D, D E, E F, F G, G H, H A; the chords of which being
joined will form the octagon A B C D E F G H A as required.

———

PROB. XIX. *In a given circle to inscribe a pentagon.* PL. 4.
FIG. 12.

Let A B C D E A be the given circle. Draw the diameters
A *f* and *g h* at right angles, cutting each other in the centre at *l:*
bisect *g l* at *i:* from *i* as a centre, with the distance *i* A, describe
an arc A *k* cutting *g h* at *k:* from A as a centre, with A *k* as a
radius, describe an arc *k* E, cutting the circumference at E: join
A E, then apply A E successively to the circumference as chords,
and A B C D E will be the pentagon required.

PRACTICAL PROBLEMS PERFORMED ON THE GROUND.

Prob. i. *To erect a perpendicular from a given point* C *to a right line* A B, *by means of a Tape or String.* Pl. 5. Fig. i.

Take two equal distances C A and C B, extend the tape to any length greater than A B, double it, put a pin in the meeting, open out the tape; place one end of the double distance, or the ring at A, and let another person hold the other end at B, and a third person take hold of the string at the pin, and stretch it out to D, then the stake at D and the point C will be in a perpendicular to A B. To illustrate this, suppose C A, C B each ten feet, then A B is twenty feet; you may extend the line to forty feet, which being doubled, the division will fall upon twenty feet, let the ring be put upon A, the divison of forty upon B; let the division of twenty feet in the middle of the line be extended out to D, while the ends A and B are held fast: then drive in the stake D, and it will give the point whence the perpendicular may be drawn to C, upon the right line A B.

N. B. Though three persons are mentioned here, one may accomplish the business by sticking an arrow in at A, and hooking the ring over it; then take a stake with two cross draughts, and drive it in at B, hook the line at forty feet round two of the cross draughts, then extend the middle at twenty as before.

———

Prob. ii. *To erect a perpendicular at or near the end of a right line*, A B, *by means of a Tape.* Pl. 5. Fig. 2.

Take any distance D B (say ten feet) extend the tape to any greater length, (say twenty feet,) fasten the ring at D, and the other end (twenty) at B, lay hold of the middle (at ten) and stretch it out to C, carry the end of the tape B round to E, until the point E be in a straight line with C and D, keeping C and D fast, and the string

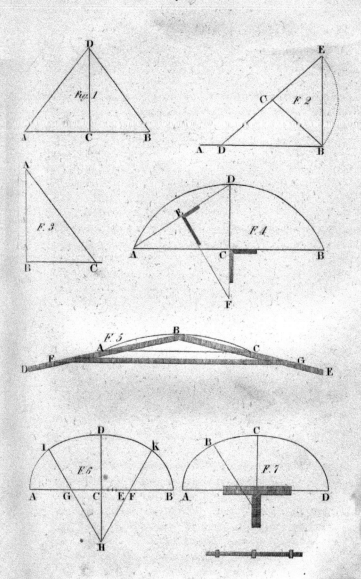

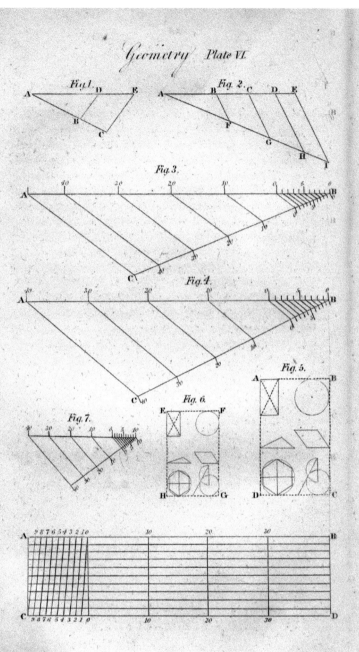

Geometry Plate VI.

Fig. 1.

Fig. 2.

Fig. 3.

Fig. 4.

Fig. 5.

Fig. 6.

Fig. 7.

N. S. Barnard, Sc.

completely stretched, drive in a stake or pin at E, then shall the points B and E be in a straight line, perpendicular to A B as required.

Prob. iii. *Another method by the Tape.* Pl. 5. Fig. 3.

Suppose the perpendicular erected upon B C from B. Take the numbers 3, 4, and 5, cr any multiple of these numbers, say, 6, 8, and 10; then 6 and 8 make 14, and 10 make 24; make B C six feet, put an arrow in at C, on which hook the ring of the tape; and fasten the division six feet at B and twenty-four feet again at C; lay hold of the line on the division fourteen feet, which carry to the point A, until both parts of the line become stretched, and the points A and B will be in a perpendicular to B C.

The same Figure.

To do the same thing by means of a *five foot rod*. Make B c three feet, with four feet; and the end of the rod resting on B, describe an arc at A, with five feet, and the end of the rod resting on C, describe another arc crossing the former at A; then shall the points A and B be in a line perpendicular to B C.

Prob. iv. *To describe the segment of a circle to any length.* A B *and perpendicular height* C D. Pl. 5. Fig. 4.

Take the middle of A B at C: fix the angle of the square at C, direct the outer edge of the stock in the straight line A B, lay a rule upon the outer edge of the blade, and draw the perpendicular D C F. In the same manner take the middle of the line
D

A D at E, and draw the perpendicular E F, the meeting Γ of the two perpendiculars will give the centre of the segment : take a slip of wood, and mark the distance D F from one end, put a brad-awl or nail through the rod at the mark, and through the point F, lay hold of the other end of the rod at D, and with a pencil at D, carry it round from A to B, pressing the pencil gently to the plane, and the point will describe the arc A B D.

N. B. Segments of circles are generally described upon a floor : but when this cannot be conveniently obtained, a temporary rough boarding is laid, which will be sufficient for brick or stone arches; but if the arc to be drawn is for joinery, and where different pieces of wood are to be fitted, the surface would require to be traversed and straightened in length and breadth.

The foregoing method may be readily applied where the space is unlimited, or the radius of a moderate length : when the radius is very great, so that a rod of sufficient length cannot be obtained, and where there is sufficient room, a wire may be used for a radius instead of a string, which cannot be depended upon in such cases, being liable to stretch ; but if you have an arc to describe, and are confined to limits, which the radius would exceed, the most eligible method will be as follows :

Fig. 5. Let A, B, C be any three points whatever, it is required to draw the arc of a circle through them without making use of the centre.

Prepare two rods, each having one of its edges straight, and each at least equal to A C the chord; lay the edge of one of the rods close to the points A and B, having one end at B; lay the straight edge of the other rod to coincide with the points B and C, having the one end also at B; notch and fix the rods together at B, and to keep the angle invariable nail a strip F G across the legs B D and B H; move the whole round, keeping the edge of the rod B D close upon the nail, pin, or brad-awl at A, and the other leg B E close to the nail, pin, or brad-awl at C ; a pencil placed at their meeting B pressing the point gently to the surface will describe the arc required.

Prob. v. *To describe a semi-elliptic arch to any length* A B *and height* C D *with a pair of compasses.* Pl. 5. Fig. 6.

Take the height C D and apply to the length from B to E towards the centre; divide the distance E C into three equal parts, set one of them towards B from E to F. Make C G equal to C F, and with the distance G F from G describe a small arc at H, and with the same distance from F describe another cutting the former arc H. Draw H G I and H F K. From the centre H with the distance H D describe the arc I K. From the centre G with the distance G I describe the arc I A. From the centre F with the same distance, or F B describe the arc K B, then A I D K B will be the semi-ellipse required.

N. B. This is a mere representation, and cannot be true; for no part of a circle is to be found in the mathematical ellipse, since the curvature is continually varying from one axis to the other. It is always lame at the junctions, and is only a make-shift, for want of better means. The following method by the trammel is correct, being derived from geometrical principles.

Fig. 7. The instrument called the trammel, consists of two pieces of wood joined together at right angles, with a groove in the middle of each; the trammel rod is a square bar with three points, or pins, made exactly to fill the grooves, and to slide easily in them, so that two of the pins must be made moveable, and to be always in a straight line with the third, which may be a pencil passing through a hole. The machine is thus prepared: set the first pin from the pencil to the height, and the second from the pencil to half the length, then put the pins in the grooves, which being fixed upon the axis, move the point B round from A to B, and describe the curve A B C D, it will be the true ellipse required.

———

Prob. vi. *Any three straight lines being given to find a fourth proportional.* Pl. 6. Fig. 1.

Let C A, A E be any two straight lines forming an angle,

Make A B equal to the first of the given lines, A C equal to the second, A D equal to the third. Join B D, and draw C E parallel to B D, cutting A E produced at E. Then will A E, be a fourth proportional to **A B, A C, A D**, or **A B, A C, A D, A E**.

PROB. VII. *To divide a line in the same proportion as another is divided.* PL. 6. FIG. 2.

Let A E be the given line, divided into the parts A B, B C, C D, D E and A 1, the line to be divided, forming any angle with A B. Join E I, and draw B F, C G, and D H, parallel to E I, cutting A 1 at F G H, then the parts A F, F G, G H, H I, will be to one another, or to the whole line A I, as the parts A B, B C, C D, D E, are to one another, or to the whole line A E.

PROB. VIII. *Any distance being given in feet and inches, of a part of one drawing to divide a given length of a similar part of another drawing into feet and inches, so as to form a proportional scale.* PL. 6. FIG. 3.

Let A B represent 57 feet 2 inches, the length of one drawing, the part between 40 and A being 7 feet 2 inches, then the distance between 40 and B will contain 50 feet; and let C B be the length of another drawing, either of greater or less extent than the former, it is required to find the scale of the new drawing. Join A C; draw 0, 0: 10 10: 20, 20: 30, 30: 40, 40, parallel to A C, cutting C B in 0: 10, 20, 30, 40; then the distance of every two adjacent divisions will be 10 fee of the new scale. The first 10 feet may be sub-divided into feet, by divisions parallel lines in the same manner, and by this means the scale of a new drawing may be found, when the whole length, or any part, and the scale of the original drawing, and the whole length, or any similar part of the required drawing are given.

PROB. IX. *A drawing being given without a scale to proportionate another, having the dimension or extent of some part of the intended drawing.* PL. 6. FIG. 4.

Draw two lines A B, B C, forming any angle A B C with each other, as before, from the angular point; on one of the lines B C set off the extent of the part of the required drawing, from B to C; from the same point B set the extent of the corresponding part of the other drawing, from B to A on the other line, and join A C. Make A B a scale of any number of divisions, as five, divide B C in the same proportion; sub-divide one of the extreme parts of A B into tenths, find the proportionate tenth of the corresponding part of B C; then will A B be a scale for the original drawing, and B C a corresponding scale for the required drawing.

Example, Figures 5, 6, 7.

Suppose A B C D A to be an original drawing, as a plate for a book, and to be of greater length or height than the page will admit of: then let the given height be E H, construct two proportional scales, fig. 7, as described in this problem, then all the dimensions and distances of the diagrams of fig. 6. will easily be proportioned to the corresponding dimensions and distances of the diagrams, fig. 5. A very accurate method, where any of the diagrams are very oblique, is to produce the sides to the boundary lines in the original drawing, then finding the corresponding points in the boundary lines of the required drawing, and by this means the angles of position may be had with the greatest correctness. In circles, the position of their centres must be found by measuring from the corresponding boundaries, and then their radii from the respective scales. Parallel lines may be drawn by the parallel ruler.

Prob. x.   *To draw a diagonal scale.*

Suppose A B to be a scale agreed upon, consisting of 50 feet, the divisions separating each two adjacent 10 feet, being 0, 10, 20, 30.   Draw the parallel lines A C, . 0, . . 10, 10 . . 20, 20. 30, 30 . . B D.   Take any convenient opening of the compass, run ten parts from A to C, and from B to D, through the divisions, draw parallels; then C D being numbered as A B: divide A 0 into 10 equal parts, and also C 0; from the points 0, 1, 2, 3, 4, &c. in A B to the points 1, 2, 3, 4, &c. draw 0, 1; 1, 2: 2, 3: 3, 4, &c.   By this means you may obtain the hundreth part of the distance A 0, or C 0, according to the parallel you measure upon; thus, suppose you required 32 feet, and 4 tenths of a foot, you must place the foot of your compass on the fourth division from 30, on the line A B, in the vertical line 30, 30, and extend the other leg along the fourth parallel, till it fall upon the diagonal 2, 3, and this extent will be equal to 32.4 feet, and thus any extent whatever may be found.

Draftsmen seldom or never make use of a diagonal scale, as persons in the habit of drawing, will judge of any small part as nearly by the eye, as if measured by the best divided diagonal scale, at least without the assistance of a glass; and thus employing a common scale will be a great saving of time.   However, in the solution of a mathematical problem in mensuration, it may be applied with advantage where time would be of less consideration, in order to obtain the accuracy desired, or to confirm the truth of a calculation.

CARPENTRY.

———

§ 1. CARPENTRY in civil architecture, is the art of employing timber in the construction of buildings.

The first operation of dividing a piece of timber into scantlings, or boards, by means of the pit saw, belongs to sawing, and is previous to any thing done in carpentry.

§ 2. The tools employed by the carpenter are a ripping saw, a hand saw, an axe, an adze, a socket chisel, a firmer chisel, a ripping chisel, an augur, a gimlet, a hammer, a mallet, a pair of pincers, and sometimes planes, but as these are not necessarily used, they are described under the head of joinery, to which they are absolutely necessary.

———

§ 3 OF SAWS.

A saw is a thin plate of steel, indented on the edge, so as to form a series of wedges, with acute angles, and for the conveniency of handling, a perforated piece of wood is fixed to one end, by means of which the utmost power of the workman may be exerted in using it.

Saws have various names, according to their use. It is obvious, in order that the saw should clear its way in the wood, that the plate should decrease in thickness from the cutting edge towards

the back, and for this purpose also, besides this additional thickness, most saws have their teeth bent towards the alternate sides
of the plate, this must always be the case where the plate is broad:
in very narrow plates the cutting edge is made thicker than usual.
Such saws as are not intended to cut into the wood their whole
breadth, have strong iron or brass backs, in order to stiffen them,
and keep them from buckling or bonding; both external and
internal angles of the teeth of saws are made to contain sixty
degrees, and the magnitude of the teeth is proportioned to the size
of the saw, and accommodated to its use.

Some saws are used for dividing the wood in the direction of
the fibre, and to any extent of distance exceeding the breadth
of the plate, at pleasure; others are only employed in cutting in
a direction perpendicular to the fibres, to any breadth or thickness;
the former case requires the front edges of their teeth to stand
almost perpendicular to the line passing through their angles, in
order to cut through, or make a way through in less time than if
set backwards, which is better adapted to the latter case: for
otherwise, the points of the teeth would run so deep into the
wood, as to prevent the workmen from pushing the saw forward
without breaking it. The saws commonly used by the carpenter,
are the ripping saw, and the hand saw; which are particularly
described under the head of joinery, as well as other saws used
in that branch.

----

§ 4. THE AXE

Is an edged tool, having a long wooden handle, for reducing
timber to a given form or surface, by paring away slices of
unequal thickness; is used by a reciprocal motion in the arc
of a circle, generally in a vertical plane, forming the surface
always in the same plane, and has therefore its cutting edge in
a longitudinal plane, passing through the handle; the slices cut

away are called chips, the operation is called chopping, and the surface reduced to its form is said to be chopped; but among woodmen the operation is called hewing.

§ 5. THE ADZE

Is also an edge tool with a long wooden handle for reducing timber to a given form of surface, by paring away thin slices of unequal thickness, by a reciprocal motion in the arc of a circle, and in a vertical plane; but its cutting edge is perpendicular to a longitudinal plane passing through the handle. It forms a much more regular and smooth surface than the axe. The operation is also called chopping.

The use of the adze is to chop and pare wood in a horizontal position.

§ 6. THE SOCKET CHISEL

Is used for cutting excavations; the lower part is a prismoid, the sides of which taper in a small degree upwards, and the edges considerably downwards: one side consists of steel and the other of iron: the under end is ground into the form of a wedge, forming the basil on the iron side, and the cutting edge on the lower end of the steel face. From the upper end of the prismoidal part rises the frustum of a hollow cone, increasing in diameter upwards; the cavity or socket contains a handle of wood of the same conic form: the axis of the handle, the hollow cone, and the middle line of the frustrum are all in the same straight line. The socket chisel, most commonly used, is about an inch and quarter or an inch and a half broad. It is chiefly used in mortising, and is the same in carpentry, as what the mortise chisel is in joinery.

Nos. 3 & 4.   E

§ 7. THE FIRMER CHISEL

Is formed in the lower part similar to the socket chisel; but
each of the edges above the prismoidal part falls into an equal
concavity, and diminishes upwards, until the substance of the
metal between the concave narrow surfaces, becomes equal in
thickness to the substance of the metal between the other two
sides, produced in a straight line, meet a protuberance projecting
equally on each side : the upper part of the protuberance is afl at,
or straight surface, from the middle of which rises a pyramid, to
which is fastened a piece of wood in the form of a frustrum of a
pyramid, tapering downwards; this piece of wood is called the
handle : the middle line of the handle, of the pyramids of the concave, and of the prismoidal parts, are all in the same straight line.

§ 8. THE RIPPING CHISEL

Is only an old socket chisel used in cutting holes in walls for
inserting plugs, and for separating wood that has been nailed together, &c.

§ 9. THE GIMLET

Is a piece of steel of a cylindric form, having a transverse handle at the upper end, and at the other end a worm or screw; and
a cylindric cavity called the cup above the screw; forming in
its transverse section, a crescent. Its use is to bore small holes;
the screw draws it forward in the wood, in the act of boring, while
it is turned round by the handle ; the angle formed by the exterior
and interior cylinders, cuts the fibres across, and the cup contains
the core of wood so cut : the gimlet is turned round by the application of the fingers, on alternate sides of the wooden lever at
the top.

§ 10. THE AUGER

Is the largest of all boring tools, it has a wooden handle at the upper end at right angles, to a long shaft of iron and steel; at the lower end is a worm or screw of a conic form, for entering the wood; so far it is similar in construction to the gimlet: the lower part of the shaft, axis, or spindle is steel, and is of a prismoidal form, to a certain distance, from the end upwards. The edges are nearly parallel, and the sides taper in a small degree upwards; the part of the shaft above the prismoid is arbitrary; but it is obvious, that in order to pass the bore freely, its transverse dimensions must be less than the lower part. The worm has its axis in the same straight line with the axis of the shaft. The lower end is hollow, or cut into a cavity on one side of the cone, and forms a projecting edge on the narrow surface of the prism called the tooth, which is brought to a cutting edge.

The part of the lower end on the other side of the cone projects before the face of the prismoidal part in the form of a wedge, the line of concourse of the two sides of the wedge forming a cutting edge. The vertex of the cone is the greatest extremity of the lower end; the cutting edge of the tooth is something higher or nearer to the handle, and the cutting edge of the wedge-like part still nearer to the handle. Any point being given as the centre of a cylindric hole on the surface of a piece of timber, the vertex of the conic screw is placed in that point; then keeping the middle line of the shaft perpendicular to, or at the inclination to be given to the surface of the timber; turn the auger round with both hands, the screw will draw it downwards into the wood, and when it has got a certain depth, the tooth will begin to cut a portion of the cylindric surface of the hole: when the part of the cylindric surface is cut half round the circumference, or perhaps a little more, the projecting wedge-like part will begin to cut out the bottom, and the core will rise in the form of a spiral shaving, by continuing to turn the handle. This construction of the auger is of very late invention, and is certainly a great improvement.

The lower part of the old form of the auger is a semi-cylinder on the outside, and the inside a less portion of a larger cylinder, the bottom of the cutting part is formed like a nose-bit: before this auger can be entered in the wood, a cavity must be first made with a gouge.

————

§ 11. THE GAUGE

Is made out of a solid piece of wood notched with an internal right angle, or consisting of two narrow planes perpendicular to each other; one of these straight surfaces forms a shoulder, the other surface has two iron teeth placed in a perpendicular to the intersection of the two surfaces, so distant from one another as to contain the thickness of the tenon, or breadth of the mortise, and the tooth next to the shoulder so far distant from the intersection, as the tenon is distant from the face. When you gauge, press the shoulder close to the wood, and the other surface of the gauge which contains the teeth, close to the other surface of the wood to be gauged; then draw and pull it backwards and forwards, and the iron teeth will scratch the wood so as to make a sharp incision or cut. When carpenters have occasion to alter their gauge for other work, they either file away the old teeth and put in new ones; or, if the distance between the old ones will answer, they cut away a parallel slice from the shoulder, or put a new piece on before it.

————

§ 12. THE LEVEL

Consists of a long rule, straight on one edge, about 10 or 12 feet in length, and another piece fixed to the other edge of the rule, perpendicular to, and in the middle of the length, and the sides of this piece in the same plane as the sides of the rule; this last piece having a straight line on one side perpendicular to the straight edge of the rule. The standing piece is generally mortised into the other, and firmly braced on each side, in order to secure it from accidents, and has its upper end kerfed in three places, one through the perpendicular line, and one on each side. The straight edge of the transverse piece has a hole or notch cut out on the under side equal on each side of the perpendicular lines. A plummet is suspended by a string from the middle kerf at the top of the standing piece, so that when hanging at length, the bottom of the plummet may not reach to the straight edge, but vibrate freely in the hole or notch. When the straight edge of the level is applied to two distant points, and the two sides placed vertically, the plummet hanging freely, and coinciding with the straight line on the standing piece, then these two points are level: but if not, let us suppose that one of the points is at the given height, the other point must be lowered or heightened according as the case may require; and the level applied each time, until the thread is brought to a coincidence with the perpendicular line. By two points, is meant two surfaces of contact, as two blocks of wood or chips, or the upper edges of two distant beams.

The use of the level in carpentry, is to lay the upper edges of joists in naked flooring horizontal, by first levelling two beams as remote from each other as the length of the level will allow; the plummet may then be taken off, and the level may be used as a straight edge. In the levelling of joists, it is best to make two remote joists level first in themselves, that is, each throughout its own length, then the two level with each other; after this, bring one end of the intermediate joists straight with the two levelled ones, then the other end of the joists in the same manner, then try the straight edge longitudinally on each intermediate joist, and such as are found to be hollow, must be furred up straight.

§ 13. TO ADJUST THE LEVEL.

Place it in its vertical situation upon two pins or blocks of wood then, if the plummet be hanging freely, and settle upon the line on

the standing piece, or if not, one end being raised, or the other
end lowered, to make it do so, turn the level end for end, and if
the plummet fall upon the line, the level is just; but if not, the
bottom edge must be shot straight, and as much taken off the one
end as you may think necessary; then trying the level first one
way and then the other as before, and if a coincidence takes place
between the thread and the line, the level is adjusted; but if not,
the operation must be repeated till it come true.

----

§ 14. THE PLUMB RULE

Is a prismatical piece of wood, with a line drawn down the middle of one of the sides, parallel to the two adjacent arrises on the
same face. Its use is to try the vertical position of posts, or
other work perpendicular to the horizon, by means of a plummet
suspended from the upper end of the rule, and a notch cut out at
the foot, in order to allow room for the plummet to vibrate freely.

In order to put up a post perpendicular to the horizon, place the
bottom of the post in its situation, and the sides as nearly vertical
as the eye may direct; if the post stands insulated, it must be fixed
in this position with temporary braces, at least from two adjoining
sides; but if very heavy, from all the four sides; then try the plumb
rule upon one side, and if the thread coincides with the line, that
side of the post is already plumb, but if not, the top must be moved
forwards or backwards, accordingly as it leans or hangs, as much
as appears to be wanted, by previously moving the front and rear
braces, and fixing them anew, while the other two remain, to stay the
other sides: apply the plumb rule again as before, and if there be a
coincidence between the line and the plummet thread, then that
face is perpendicular, but if not, the several similar operations must
be repeated till found to be so. Proceed in the same manner with
the other two parallel sides of the post, until they are made plumb,
and by this means the post will be set in a true vertical position.

CARPENTRY.

§ 15. THE HAMMER

Consists of a piece of steel, through which passes a wooden handle perpendicularly; the steel is flat at one end, or in a small degree convex. The use of the hammer is for driving nails into wood by percussive force. The other end of the hammer, that is not used for driving nails, is sometimes made with claws, and sometimes with a rounded edge, like a semi-cylinder. The claws are for laying fast hold of the head of a nail, to be drawn out of a piece of wood; for this purpose the back of the hammer is rounded, so that the hammer, in the act of drawing the nail, may not penetrate with its other extremity into the wood; and this also lessens the distance of the force to be overcome from the fulcrum, and consequently increases the power employed. When the hammer is used, place the back of it upon the wood, and the claws so as to have the nail fast between them, lay hold of the handle and pull the contrary way to that side of it on which the nail is; then, if the force be sufficient, the nail will be drawn out of the wood, and the nail thus drawn will come out almost straight. Some people, instead of pulling the handle of the hammer the contrary way to the side on which the nail is on, (and thereby making it describe a circle in a plane, perpendicular to the surface of the wood, and through the longitudinal direction of the head,) turn the hammer sideways: the nail is easier drawn by this way, but then the surface of the wood is more injured, as well as the nail, which is frequently so much bent as not to be of any more use. Claw hammers are chiefly used in the country; and those with their other extremity rounded like a cylinder, are used in town for clinching and rivetting. In driving a nail, when the hammer comes in contact with the head of the nail, if the striking surface is not perpendicular to the shank of the nail, the nail will not be driven into the wood, or only in a small degree, but will be bent sideways towards an oblique angle, and will thus frequently break the nail, unless it be well entered, and so strong as to resist the force acting thus obliquely. The reader must here observe, that no force can act with its full effect upon another, unless in a line perpendicular to the surface of contact.

§ 16. THE MALLET

Is similar in its construction to the hammer, but the head is a thick block of wood, of a structure in form of the frustum of a pyramid, the side of this frustum tending to some point in the handle continued. Its use is for mortising and driving pins into wood. The object is struck by the narrow sides of the mallet.

---

§ 17. THE BEETLE OR MAUL

Is a large mallet to knock the corners of framed work, and to set it in its proper position, and is sometimes used for driving short piles into the ground, where it would be unnecessary to use greater power. The handle is about three feet in length, and for these heavy purposes both hands are employed. This is more used in the country than in London, where they use a sledge hammer for the same purpose.

---

§ 18. THE CROW

Is a large bar of iron, used as a lever to lift up the ends of heavy timber, in order to lay another piece of timber, or a roller, under it. One end of the crow has claws.

---

§ 19. THE TEN FOOT ROD

Is a rod about an inch square, divided in its length into feet and inches, for the purpose of setting out work. The method of raising a perpendicular by a ten foot rod, is described in the Practical Geometry, page 25. PROB. III. Instead of a ten foot rod, some use two five foot rods for the same purpose.

§ 20. HOOK PIN

Is a conical piece of iron, with a hooked head, declining upwards in the form of a wedge. The top is flat, for the purpose of driving it down; and the shoulder which rises from the cone, stands perpendicular to the axis, and is used for driving it out of a hole, when it is fixed fast. The hook pins are the same in carpentry, as what the draw bore pins are in joinery, viz. they are employed after the tenons have been entered in the mortise and bored, as shall be presently shown, in drawing the shoulders of the tenons home to their abutments in the mortise cheeks : when there are several mortises and tenons in the same frame, as many hook pins are employed. The method of boring, and using the hook pins, is thus : bore a hole first through the mortise cheeks, not very distant from the abutments ; enter the tenon, and force it home to its shoulders as near as you can ; mark the tenon by the hole, and draw the tenon out of the mortise. Then pierce a hole through the tenon, about one third of its diameter nearer to the shoulder, and enter the tenon again, bringing the shoulder as near to its abutment as possible ; drive in the hook pin with considerable force ; the convex circumference will bear upon alternate sides of the mortise and tenon, viz. upon the farther side of the hole of the tenon, and upon the nearest side of the mortise from the joint ; the shoulder of the tenon being brought home to its abutment, the hook pin may be drawn out of the hole ; for this purpose there is a hole through the upper part of it, by which it is sometimes drawn out with another hook pin ; but if driven in very fast, it will require the assistance of a hammer to strike it upon the shoulder upwards, and two or three smart blows will soon loosen it ; when drawn out, enter the pin, and drive it home with force, or till it be sufficiently through and fast, so as not to be driven farther without breaking.

F F

§ 21. THE CARPENTER'S SQUARE

Is a square of which both stock and blade consists of an iron plate of one piece ; it is in size and construction thus : one leg is eighteen inches in length, numbered from the exterior angle, the bottom of the figures are adjacent to the interior edge of the square, and consequently their tops to the exterior edge : the other leg is twelve inches in length, and numbered from the extremity towards the angle ; the figures are read from the internal angle, as in the other side ; each of the legs are about an inch broad. This implement is not only used as a square, but it is also used as a level, and likewise as a rule : its application as a square and as a rule is so easy as not to require any example : but its use as a level, in taking angles, may be thus illustrated ; suppose it were required to take the angle which the heel of a rafter makes with the back, apply the end of the short leg of the square to the heel point of the rafter, and the edge of the square, level across the plate, extend a line from the ridge to the heel point, and where this line cuts the perpendicular leg of the square, mark the inches, and this will show how far it deviates from the square in twelve inches.

----

§ 22. OPERATIONS.

Having now mentioned the principal tools, and their application, it will here be proper to say something of the operations of Carpentry, which may be considered under two general heads ; one of individual pieces, the other the combination of two or more pieces.

Individual pieces undergo various operations as sawing, planing, rebating, and grooving, or ploughing : the operation of the pit saw is so well known as hardly to need a description ; planing, rebating, grooving, or ploughing, are more frequently employed in Joinery, and will be there fully described. The other general head may be sub-divided into two others, viz. that of joining one piece of timber to another, in order to make one, two, or four angles, the other that of fastening two or more pieces together, in order to form one piece, which could not be got sufficiently large or long in a single piece; there are two methods of joining pieces at an angle, one by notching, the other by mortise and tenon.

Notching is the most common and simple form that prevails in permanent works, and in some cases the strongest for joining two pieces of timber together, at one, two, or four angles : the form of the joint in this is varied according to the situation, the positions of the sides of the pieces, the number of angles, the position of the pieces, and the quantity and direction of the force impressed on one or both pieces, or according to any combination of those circumstances. The most useful are the following.

———

§ 23. *To join two pieces which are to form four angles, and the surfaces of one piece are both parallel and perpendicular to those of the other.*

A notch may be cut out of one piece, the breadth of the other, which may be let down on the first piece, or the two pieces may be reciprocally notched to each other, and for further security, nails, spikes, or pins, may be driven through both : this form is applicable where each of the pieces are equally exposed to strain in any direction : when one piece has to support the other transversely, the upper piece may have a notch cut across it to a breadth; suppose two-thirds of the thickness of the piece below, and the lower piece must have an equal notch cut out on each upper arris, leaving two-thirds of the breadth of the middle entire, by which the strength of the supporting, or lower piece, is less diminished than if a notch of much less depth had been cut the whole breadth : this mode is applicable to carcass roofing, in letting the purloins down upon the principal rafters, and the common rafters again upon these; also in carcass flooring, it is employed in letting down the bridging joists upon the binding joists.

§ 24. *To join one piece of Timber to another, to form two right angles with each other, and the surfaces of the one to be parallel and perpendicular to those of the other, and to be quite immoveable, when the standing piece is pulled in a direction of its length, while the cross piece is held still.*

Dovetail the end of the perpendicular piece, that is, form it like a truncated isosceles triangle, the wide part being on the extremity, make a corresponding reverse in the other, and if both these pieces be horizontal, and the former laid upon the latter, they will answer the intended purpose without the addition of nails, spikes, or pins: in this mode, if the timber is not sufficiently seasoned, the perpendicular piece may be drawn out of the transverse piece, to a certain distance, according to the degree of shrinking.

———

§ 25. *Another mode,*

Which prevents the perpendicular piece from being drawn out of the transverse piece, allowing that the timber should shrink, is to notch the transverse piece, so as that, if the breadth be supposed to be divided into five equal parts, and three of these be notched from one edge, and one from the other, leaving one part entire, observing that these two notches should not be cut more than one third of the thickness through ; then cut a notch out of the perpendicular, to fit the entire part of the transverse, leaving two-fifths entire towards the extremity, and when the two pieces are joined together, the notch and the entire part of the perpendicular piece will respectively fit the entire part, and the broad notch of two-fifths of the transverse piece. If the upper piece press upon the under piece, by its own weight, or with an additional force, neither nails, spikes, nor pins will be necessary.

These methods of framing a piece of timber, at right angles to another, are used in cocking down the beams of a building upon the wall-plate ; but the latter method is more generally employed than the former, as being more perfect ; either method is infinitely superior to mortise and tenon for such purpose.

——

§ 26. *To notch one piece of Timber to another, or join the two, so as to form one right angle, in order that they may be equally strong, in respect to each other.*

Notch each piece half through, and nail, spike, or pin them together ; or they may be partly notched on each other, and the inner edge of one again notched, leaving the substance sufficiently thick below each notch, and a part entire at the inner edge ; cut the corresponding reverses in the other piece, and when the two are joined, neither can be drawn out of the other : these two methods of joining a piece of timber to form a right angle with another, are applied to wall-plates and bond timbers at the corners of a building ; but wherever the thickness of the walls will admit, it is much better to make the end of each piece to pass the breadth of the other as much as possible, so that by this means four right angles will be formed instead of one ; then the two may be equally notched as in the former case

——

§ 27. *To fix one piece of Timber to another, forming two oblique angles, so that the standing piece cannot be drawn out of the transverse.*

Cut a dovetail notch in the transverse piece, keeping the edge straight upon the side next to the obtuse angle, that is, forming the dovetail on the side of the acute angle ; make the corresponding notch upon the piece which has the two angles on the same side, and nail, spike, or pin them together if necessary : this form is particularly applicable to roofing.

§ 28. *To cut a rebated notch in the end of a Scantling, or piece of Wood.*

If the piece is not above three or four inches in either dimension, it may be cross-cut with the hand-saw to the depth, and the piece may be cut longitudinally out, or in the direction of the fibres with the same : but if the stuff is very broad, as a plank or board, and the notch is to be cut in the breadth of the board, then you may cross-cut the face with the hand-saw as before, and cut the piece out with the adze to the depth required; if it is to be cut from the edge of a board or plank, you may proceed as at first with the hand-saw only.

———

§ 29. *To cut a grooved notch, or socket in a piece of Timber.*

Cross-cut the two ends or sides with the hand-saw to the intended depth; then, if the notch is sufficiently long or broad to admit of the breadth of the blade of the adze, you may cut out the wood between the two kerfs with the adze ; but if the width or breadth of the tenoned piece is not of sufficient extent, you may then cut out the intermediate wood between the kerfs with the socket chisel, and smooth the bottom of the notch with the paring chisel.

———

§ 30. *To cut a Tenon.*

This operation is only a double rebated notch; and consequently the methods for cutting the tenon are the same under like circumstances of size and dimensions. See also the next article.

———

§ 31. *To frame one Timber at right angles to, and at some distance from, either end of another, both pieces being of the same quality.*

To do this, the piece of timber which is to stand perpendicular to the other, must be reduced of its thickness by cutting away two rectangular prisms from both ends, and leaving another rectangular prism in the middle of the thickness, commonly called a tenon, which is made to fit a corresponding excavation, called a mortise, taken out of the other piece, so that when both pieces are joined together, two of the surfaces of the one piece will be straight with two of the surfaces of the other, and the other two remaining surfaces of the one piece will be perpendicular to the other two remaining surfaces of the other; and if properly joined, the superfices of both pieces will come in contact with each other, so as to leave no interstice or cavity.

Before the mortise and tenon is made, it will be proper to say something of the proportion between the thickness of the tenon, or breadth of the mortise, and the thickness of the stuff: Suppose the tenon to be entered in the mortise, and driven home; and suppose the piece which has the mortise, to be held still, while a force is applied to the other end of the tenoned piece, so as to act transversely to the mortised piece, then one or other must give way. It is evident that if the mortise cheeks are too thin, they will split, or if the tenon be too thin, it will break transversely; there is, therefore, some proportion between the breadth of the mortise and the thickness of the stuff, so that the one shall be equally strong with the other, to resist this kind of strain. Another thing which will affect this proportion, is, whether the junction is to be supported, as in wall-plates, or unsupported, as in joisting; a thinner tenon will be required if unsupported, than if supported; for suppose that the junction has no support, the surface of both parts lying horizontally; and suppose a weight or force upon the tenoned piece, near to the shoulder, pressing vertically downwards, while the mortised piece is fixed at both ends, and the tenoned piece is also fixed at its remote end; likewise suppose that the width of the mortise is one third of the thickness of the stuff, it will perhaps be found that the under cheek of the mortise will split away, while the tenon will remain unbroken; the mortise, therefore, requires to be still less: but there is another reason, equally powerful, which corroborates this practice, which is, that by cut-ting away one third of the substance, the mortised piece woula be weakened too much when thus unsupported, as is the case in joisting.   Though we cannot determine with mathematical accu-racy, nor by any result of experiments, common practice has sanctioned the thickness of the tenon to be about one fifth of the thickness of stuff; this being fixed, we shall now proceed to the practice.

First square the shoulder, by drawing three lines, one perpen-dicular to the thickness of the tenon, and each of the other two to meet this line perpendicular to the adjoining arrises, on which the first line was drawn; then mark the breadth of the tenon, at the place where the mortise is to be cut, in the length of the mortised piece; through each extremity draw a line by the iron square, perpendicular to the arrises on the one side on which the mortise is to be cut, and at the intersection of the lines, with one of the adjoining arrises, draw two other lines on the con᠁ ᠁us side: then, where each of these lines meet the other arris, ᠁᠁᠁ lines in the same manner upon the third side; so that each of the three contiguous sides will have two lines at right angles to the arrises of that side.   Take the gauge, described in section 11, and gauge the tenon from the face, and the mortise from the same side, which is to be flush with it.   Then entering the handsaw by the lines drawn on the shoulder, cut the shoulders to the gauge lines, and saw off the tenon cheeks, and thus you have the tenon completed. Then with the socket chisel and mallet knock out the core of the mortise; then drawbore your work together with the hook pins as in section 20, and the work will be completed.

---

§ 32.   *To join two Timbers by Mortise and Tenon, at a right angle,
   so that the one shall not pass the breadth of the other.*

Let us suppose that each of the pieces to be framed are of yel-low fir, or both of the same quality of wood.   It is evident, that it the mortise were cut away the whole breadth of the tenon, and the tenon of the same breadth as the piece it is formed on, that the one could not make any resistance to the other without the assistance of a pin. In order to accomplish this, the mortise must not be cut to its full breadth, but must want a certain part of that towards the end of the tenoned piece; our next inquiry must be the proportion between the length of the mortise, and breadth of the tenoned piece, as it must be considered the strain which the mortise is liable to, is splitting, and that of the tenon, is in breaking transversely to the fibres; for there is a certain proportion between the breadth of the tenon, and breadth of the piece on which it is cut, so that the one will resist equally the other This is a point that has not been mathematically ascertained; however, common practice allows the tenon to be reduced about one third of its breadth, and consequently the breadth of the tenon two thirds, and the length of the mortise two thirds also. As to the thickness of the tenon, or breadth of the mortise, it is the same as we mentioned in the preceding case, and will differ according as it is to lie hollow, or lie upon a solid. The cutting of the tenon, and taking out of the mortise, is the same as has been shewn in the preceding case, the pinning the same as in section 20.

----

§ 33 *Of Foundations and Timbers, in joisting and walling.*

The foundations being excavated to the intended depth, the ground must be examined, by trying whether it is sufficiently firm in all places, so as to support the weight of the intended building. There are several means of securing foundations without piling, should any artificial means be required; but as our present subject is carpentry, and as these do not come under the carpenter's profession, we will first suppose that the intended building is to be brick or stone, and that the foundation is infirm, piles must then be prepared, such, that their thickness may be about a twelfth part of their length. The distances which those piles will require to be

Nos. 3 & 4. F

disposed, and the momentum required to drive them, will depend
on the weight of the building; for the weight of the ram used in
driving them, ought not to be more than what would be sufficient
for the purpose, as a greater number of men, or power, would
need to be employed, which would occasion an unnecessary expense. We will now suppose the piling to be completed, so as to
be sufficient for supporting the intended building; some people lay
a level row of cross bearers, called sleepers, and plank above;
but then observe, before the planking is laid, that all the interstices
should be levelled up to the top of the sleepers, with bricks, &c.
The planking, however, will not be necessary, provided that the
piling be sufficiently attended to, and thus the expense of the foundation will be materially lessened. All timber whatever, of which
the thickness stands vertical in the building, being liable to shrink,
will also make the building liable to crack, or split, at the junctions
with the return parts. In cases where the ground is not very soft,
a balk is sometimes slit in halves, and these either laid immediately
at the bottom, or at the height of two or three courses, and this
will frequently prevent settlements, which are occasioned by an
unequal pressure of the piers, and the intermediate brick-work or
masonry, under apertures. Suppose the foundation to be brought
up to its height, or to the level of the under sides of the ground
joists; the ground plates must be laid, and sleepers, at eight or ten
feet distance where the floors are intended to be boarded: these
sleepers are supported upon small pillars or piles of brick, or by
stones, at five, six, or eight feet distance, according to the substance of timber used for the sleepers, and their ends supported by
the walls. The next thing is to lay the ground joists. When the
bricklayer has got to the top of the first windows, the carpenter
may lintel the windows: but if the joisting of the next floor is laid
upon the lintels, the wall-plate and the lintels will form one continued length of timber, which will be much stronger than lintels,
having only nine or ten inches bearing upon the walls. Suppose
now the wall-plates laid round the exterior walls, and returned in
flank or party-walls, except at the flues, and likewise laid in cross-

walls of brick or stone; or if a timber partition is required, and the joisting to be supported by this partition, the partition is seldom carried up, the joisting is first laid and levelled; instead of the partition, a plank or other piece of timber is laid under the joisting at the place, and this supported by uprights, which are forced up with wodges, so as to bring the top of the joists to a level; before the joisting is put down, the trimmers of stairs and chimnies must be framed in. If a double floor is to be laid with girders, be sure to lay templets, or short pieces of timber, under the girders, as this will distribute the pressure over a greater surface, and thereby prevent settlements. The naked flooring being laid, in carrying up the second story, bond timbers must be introduced opposite to all horizontal mouldings, as bases and surbases. It is also customary to put a row of bond-timber in the middle of the story, of greater strength than those for the bases and surbases. The work being so far advanced, we will suppose the building roofed in and completed; as there will be immediate occasion for resuming the subject in the description of a wooden building.

§ 34. *Stud-work, and Plaster Buildings.*

The foundation being made secure, and the several scantlings for ground-plates, principal posts, posts, bressummers, girders, trimmers, joists, &c. being prepared and framed, agreeable to their several situations. Timbers laid on the foundation, or next to the ground, are generally of oak, as ground-plates, which should be about eight inches broad, and six inches vertically. The front and rear plates are to be framed by mortise and tenon, the front and rear plates being mortised, and the flank pieces consequently tenoned. Sometimes the flank pieces are mortised to receive the joists. The ground plates are to be bored with an inch and half auger, and pinned together with oak pins, made taper towards the point, and so strong as to withstand the blows of the mallet, when driven tight into the hole. As the wood which carpenters work upon is generally heavy timbers, a block is laid under the corner to bear the plate off the foundation, so as to allow room for driving of the hook pins ; when the wooden pins are driven, remove the blocks, and let the plates bed firmly on the foundation. But before the pins are driven, if there be any girders it must be fitted in, and all the joisting and trimmers, for they cannot be got in afterwards. We shall suppose that every thing is got to its birth, and the work pinned together. Four corner posts, eight inches by six, viz. of the same scantling as the ground plates, are erected, presenting their narrow sides to the front, and extending the whole height of the building, till they meet the wall-plates. These corner posts are called principal posts, and are mortised and tenoned into the ground-plates, and also for the purpose of being inserted into the rising-plates. At the height of the principal story, two mortises must be cut in each principal post ; which being set up, enter the tenons of the next bressummers into the mortises, and stay the principal posts, by means of temporary braces, fixed to the framed work of the floor. Set up the several intermediate story posts, or those which are framed into the interstices, and tenon the ends of these posts into the bressummers or interstices, as it may happen whether there are interstices between the bressummers or not. Proceed in like manner with the bressummers, girder, and joists, of the next story. It does not always happen that there is a girder, but if one side of it should prove to be wainy, that side must be turned upwards, and the shoulders of the joists must be scribed upon the wains.

We shall now suppose, the principal posts, story posts, or other intermediate posts, bressummers, girders, floor joists, trimmers, and trimming joists, all completely fitted together, you may proceed to pin the work together, and put on the raising plates, which are let down upon the tenons of the principal posts, and then complete the roof ; you may then begin to put up the truss partitions, if there be such, and fill in the larger interstices in the outside framing, and in these partitions with quarters.

§ 35. What now remains to be done belongs to the joiner, and will therefore be found under the article Joinery.

In the description of this wooden fabric, as there are several particulars respecting the scantlings and bearings of timbers, not mentioned, the following table may be referred to, not only to sup ply these wants, but on various other occasions.

In the following tables, the first vertical column contains the heights or bearings in the clear of timbers; the second, the scantlings in inches for firwood; and the third, the scantlings in inches for oak wood; the corresponding parts are to be found in each horizontal row, as is sufficiently plain from the tables.

§ 36. TABLE I.

BEARING POSTS.

Height.	Fir.	Oak.
Feet.	Inches by Inches.	Inches by Inches.
8	6 × 10	7 × 12
10	7 × 11	8 × 13
12	8 × 12	9 × 14
14	9 × 13	10 × 15
16	10 × 14	11 × 16
18	11 × 15	12 × 17
20	12 × 16	13 × 18

§ 37. The table of bearing posts here given, is considered as sufficient only for supporting two or three stories of a dwelling house, it is impossible to give a table that will be adequate to every class of building. These scantlings do not depend upon the height of the building, but upon the weight with which the several floors are loaded.

The supporting timbers required for the construction of a ware-house, ought to be very different from those employed in a common dwelling house. It must be farther observed that all bearing

F 2

posts which stand insulated, ought to be exactly square; but, as in general they are stayed sideways by doors, windows, or intertices; the sides of the pieces employed are of unequal dimensions: giving a greater depth, requires less timber to make them equally strong, and by making them thinner, gives more ample area for light, which is particularly wanted in shop stories. Another observation; the table above is not constructed, so as to make the story posts at different heights equally strong, even under the same circumstances of weight, as higher posts would be more liable to accidents than lower ones, so that there is a continued increase of strength from the lower to the higher posts. We cannot say positively, what the exact scantlings for bearing posts of given heights ought to be, though the weight which they have to support were known, as we have no detail of experiments sufficient to enable us to establish a principle of calculation. We have therefore, nothing else to depend upon but our experience, and what we see commonly put in practice. Two practical men will not always exactly agree, in what ought to be a standard under particular circumstances. The breaking of timber by compression, is so intricate of itself, that men of science have not agreed as to the general law by which a transverse fracture is produced. With regard to the difference of strength between fir and oak, Muchenbroek asserts, on the authority of his own experiments, that although oak will suspend half as much again as fir, it will not support as a pillar, two thirds of the load: upon this authority also, the author has ventured to make the oak scantling larger than the fir.

§ 38. TABLE II.

GIRDERS.

Bearing.	Fir.		Oak.	
Feet.	Inches by Inches.		Inches by Inches.	
12	10 × 8		9 × 7	
16	12 × 10		11 × 9	
23	14 × 12		13 × 11	
24	16 × 14		15 × 13	

§ 39. TABLE III.

BRIDGING JOISTS.

Bearing.	Fir.		Oak.	
Feet.	Inches by Inches.		Inches by Inches.	
4	4 × $2\frac{1}{2}$		$3\frac{1}{2}$ × $2\frac{1}{4}$	
6	5 × $2\frac{1}{4}$		$4\frac{1}{2}$ × $2\frac{1}{4}$	
8	6 × $2\frac{1}{2}$		$5\frac{1}{2}$ × $2\frac{1}{2}$	
10	7 × $2\frac{1}{4}$		$6\frac{1}{2}$ × $2\frac{1}{4}$	

§ 40. TABLE IV.

BINDING JOISTS.

Bearing.	Fir.	Oak.
Feet.	Inches by Inches.	Inches by Inches.
8	7 × 4	6 × 4
10	8 × 4	7 × 4
12	9 × 4	8 × 4
14	10 × 4	9 × 4

§ 41. TABLE V.

TIE BEAMS.

Bearing.	Fir.	Oak.
Feet.	Inches by Inches.	Inches by Inches.
20	8 × 4	7 × $3\frac{1}{2}$
30	10 × 6	9 × $5\frac{1}{2}$
40	12 × 8	11 × $7\frac{1}{2}$
50	14 × 10	13 × $9\frac{1}{2}$
60	16 × 12	15 × $11\frac{1}{2}$

§ 42. TABLE VI.

PRINCIPAL RAFTERS.		
Bearing.	Fir.	Oak.
Feet.	Inches by Inches.	Inches by Inches.
12	5 × 3	6½ × 3½
18	6½ × 4	7½ × 4½
24	8 × 5	9½ × 5½
30	9½ × 6	10½ × 6½
36	11 × 7	12½ × 7½

§ 43. TABLE VII.

PURLINES.		
Bearing.	Fir.	Oak.
Feet.	Inches by Inches.	Inches by Inches.
6	7 × 4	6½ × 3½
8	8 × 5	7½ × 4½
10	9 × 6	8½ × 5½
12	10 × 7	9½ × 6½
14	11 × 8	10½ × 7½

§ 44. In table VI. As principal rafters are always in a state of compression, the oak scantlings are increased according to the aforesaid experiments. All ties should therefore be made of oak, and all compressed or straining pieces of fir.

§ 45. TABLE VIII.

SMALL RAFTERS.		
Bearing.	Fir.	Oak.
Feet.	Inches by Inches.	Inches by Inches.
8	$4\frac{1}{2} \times 2\frac{1}{2}$	$4 \times 2\frac{1}{4}$
10	$6 \times 2\frac{1}{3}$	$5\frac{1}{2} \times 2\frac{1}{4}$
12	$7\frac{1}{2} \times 2\frac{1}{2}$	$7 \times 2\frac{1}{4}$

All beams ought to be cut or forced to a camber, an inch for every 20 feet : as all framed work will shrink, and sag after being put together.

Roofs are much stronger when the purlines run above the principal, than when framed in.

In all case or tail bays, in floors or roofs, the bearings of either joists or rafters, ought not to exceed 12 feet.

————

Abstract of the Building Act, as far as regards the Carpenter, 14 Geo. III. *which refers only to London, and the several Parishes within the Bills of Mortality*

Those timber partitions between building and building, that were erected, or begun to be erected before the passing of the act, may remain till one of the adjoining houses is rebuilt, or till one of the fronts, or two thirds of such fronts, which abut on such timber partition, is taken down to the bressummer, or one pair of stairs floor, and rebuilt.

Proprietor of a house or ground to give three months notice to pull down such wooden partitions when decayed, or of insufficient thickness, and to be left with the owner or occupier of such a house, and if empty, such notice to be stuck up, in and on the front door, or front of such house.

No timber hereafter to be laid in any party arch, nor in any party wall, except for bond to the same ; nor any bond timber, within 9 inches of the opening of a chimney, nor within 5 inches of the flue, nor any timber within 2 feet of any oven, stove, copper, still, boiler, or furnace.

All framed work of wood for chimney breasts, to be fastened to the said breast with iron work as hold fasts, wall hooks, spikes, nails, &c. nor driven more than 3 inches into the wall, nor nearer than 4 inches to the inside of the opening of the chimney.

No timber bearer to wooden stairs let into an old party wall, must come nearer than 8 and a half inches to the flue, nor nearer than 4 inches to the internal finishing of the adjoining building.

No timber to be laid under any hearth to a chimney, nearer than 18 inches to the upper surface of such hearth.

No timber must be laid nearer than 18 inches to any door of communication through party walls, through warehouses or stables.

Bressummers, story posts, and plates thereto, are only permitted in the ground story, and may stand fair with the outside of the wall, but must go no deeper than 2 inches into a party wall, nor nearer than 7 inches to the centre of a party wall, where it is two bricks thick, nor nearer than 4 inches and a half, provided the party wall does not exceed one brick and half in thickness.

Every corner story post must be of oak, at least 12 inches square, when employed for the support of two fronts.

Window frames and door frames to the first, second, third, and fourth rate classes, are to be recessed in reveals, 4 inches at least.

Doorcases and doors to warehouses only of the first, second, third or fourth rate classes may stand fair with the outward face of the wall.

No external decoration to be of wood, except cornices or dressings to shop windows, frontispieces to door-ways of the second, third, and fourth rate classes, covered ways or porticos to buildings; but not to project beyond the original line of the house in any street or way; such covered way or portico not to be covered with wood.

Nor such cornice, covered way, or the roof of portico to be higher than the under side of the cill to the windows of the one pair of stairs floor.

No flat gutter or roof, nor any turret dormer, or lanthorn light, or other erection placed on the flat of the roof belonging to the first, second, third, fourth, and fifth rate classes to be of wood or timber

No wooden water trunks must be higher from the ground, than the tops of the windows of the ground story.

———

PLATE VII.

Fig. 1 the axe used in chopping timber by a reciprocal circular motion, generally in a vertical plane, and with the cutting edge in that plane.

Fig. 2 the adze used also in chopping timber by a reciprocal motion, generally in a vertical plane, but with the cutting edge perpendicular to the plane, and thereby forming a horizontal surface.

Fig. 3 the socket chisel used in mortising; it must be observed, that the socket chisel is not always the breadth of the mortise, but generally less, particularly when the mortise is very wide.

Fig. 4 mortise and tenon guage.

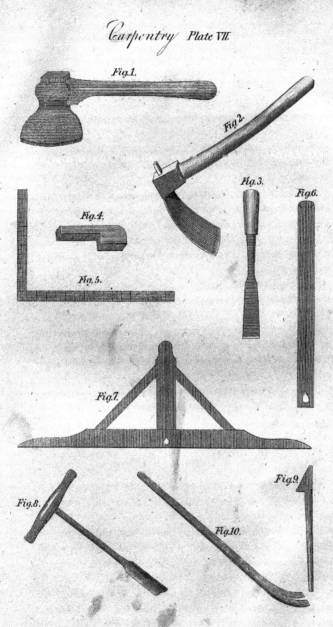

Carpentry Plate VII.

Fig.1.

Fig.2.

Fig.3.

Fig.6.

Fig.4.

Fig.5.

Fig.7.

Fig.8.

Fig.9.

Fig.10.

Carpentry. Plate VIII.

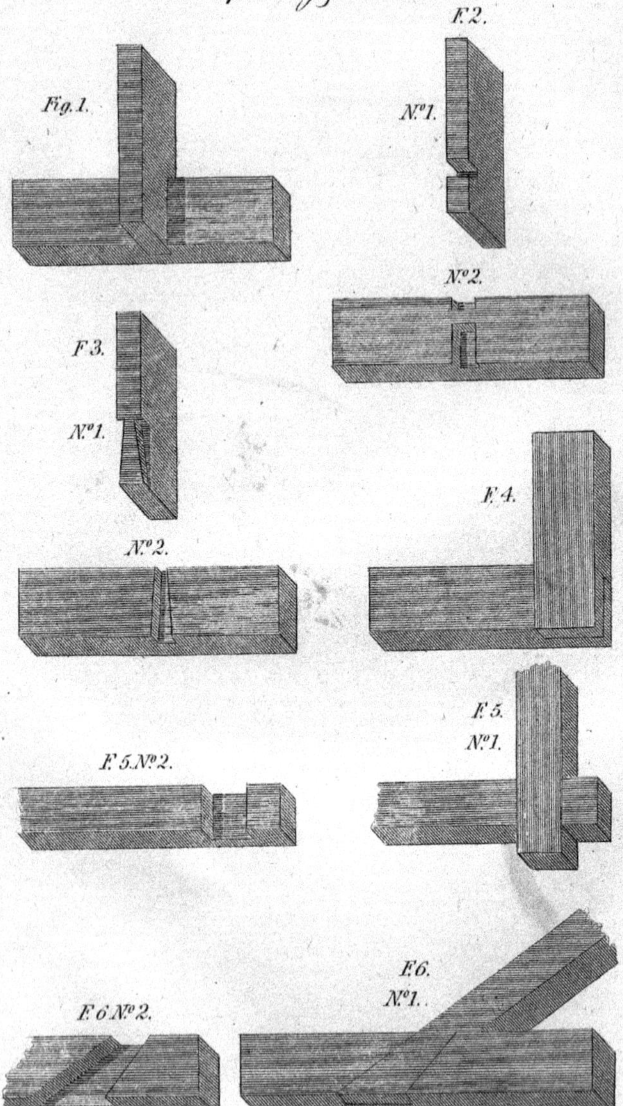

Fig. 1.

F. 2.
Nº 1.

Nº 2.

F 3.
Nº 1.

Nº 2.

F. 4.

F. 5. Nº 2.

F. 5.
Nº 1.

F. 6.
Nº 1.

F. 6 Nº 2.

Fig. 5 the carpenters' square.

Fig. 6 the plumb rule.

Fig. 7 the level.

Fig. 8 the auger.

Fig. 9 a hook pin for drawboring.

Fig. 10 the crow.

PLATE VIII.

Fig. 1 the manner of cocking tie beams with the wall plates fitted together. See § 25.

Fig. 2 shows the manner by which the cocking joint is fitted together, No. 1 part of the end of the tie beam, with the notch to receive the part between the notches in No. 2, which is a part of the wall plates. See § 25.

Fig. 3 dove-tail cocking; No. 1 the male or exterior dove-tail cut out on the end of the tie beam: No. 2 the female or interior dove-tail cut out of the wall plate, to receive the male dove-tail. See § 24.

Fig. 4 the manner of joining two pieces together to form a right angle, so that each piece will only be extended on one side of the other, by halving the pieces together, or taking a notch out of each, half the thickness. See § 26.

Fig. 5 two pieces joined together, forming four right angles, when one piece only exceeds the breadth of the other by a very short distance: No. 2 the socket of one piece, which receives the neck or substance of the other. This and the preceding are both employed in joining wall plates at the angle; but the latter is preferable, when the thickness of walls will admit of it.

Fig. 6 the method of fixing angle ties: No. 1 part of angle tie, with part of the wall plate: No. 2 the wall plate, showing the socket or female dove-tail. Though the angle tie is here shown flush with the wall, in order to show the manner of connecting the two pieces together; the angle tie is seldom, or never let down flush, as this would not only weaken the angle tie, but also the plate into which it is framed. See § 27.

PLATE IX.

Fig. 1 plan of a floor where the joists would have too great a
bearing without a girder, and where the walls in the middle of the
apartment are perforated with windows below.   If there were no
windows, the place of the girder would be obviously in the middle
of the wall, in order to make the strongest floor out of timber of
given scantlings, or to make it equally strong with the least quantity of timber; but as there is an opening, and if the end of the
girder were to be laid over that opening, it would render the walls
liable to fracture, which would be still a greater error than the former; to avoid this evil, the girder must then lie upon a solid pier,
and to make the best of this circumstance, so as to be at the least
expense in timber, or to make the strongest floor out of given timbers, the end of the girder must be placed as near to the aperture
as possible, so as to have a solid bearing, and the other end as far
distant from the middle line, upon the alternate side of this line :
and thus the middle of the girder would still be in the middle of
the length.   Some objections may be raised against this method of
placing the girder, as it only divides the centre joists equally; but
the answer to this is, that the greatest stress upon the floor is always in the middle : and therefore, as the joists are equally divided
in the middle, there is the greatest strength where there is most
occasion for it; and likewise, taking all circumstances together,
the middle is not capable of sustaining the same weight as other
parts of the floor nearer to the extremes are : however, it still remains as a question, whether a girder placed in this position, or
stronger joists running the other way, would make the cheapest
floor : this I shall leave, as circumstances in practice may determine.

Fig. 1. *Explanation of the Timbers in a single Floor.*

A, A, A, &c. plan of walls.
B, B, B, the flues of chimnies.

Carpentry Plate IX.

Fig.1.

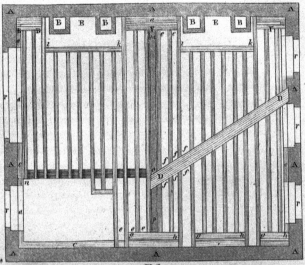

F.2.

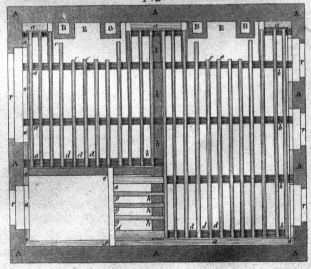

C, C, C, the upper side of wall plates.

D, D, girder.

E, E, fire-places.

e f, e f, e f, &c. tail bays of joists framed into girder.

g h, g h, g h, tail trimmers framed into trimming joists, in order to prevent the ends of the timbers as much as possible from going into the wall, according to the Building Act.

i k, i k, hearth trimmers.

m o a quarter partition between rooms.

n o p a nine inch wall, inclosing stairs.

———

Fig. 2. *Explanation of the Timbers in a double Flóor.*

In this, the plans of the walls, flues of chimneys, and upper side of wall plates are denoted by the same letters as the same things in the preceding explanation are. The other parts are as follow :

a b, a b, a b, binding joists.

c d, c d, c d, &c. bridging joists.

e f, stair trimmer.

g h, single joists framed into stair trimmer.

It may be proper here to observe, in this explanation, that any row or compartment or joisting to which the flooring boards are attached, whether in a double or single floor, between any two adjacent supports, is called a bay of joisting; a bay of joisting next to the wall, is called a tail bay : and those between two girders, or between two binding joists, are called case bays: thus in fig. 1 the joisting on either side of the girder is called a tail bay: and in fig. 2 there are two case bays, and two tail bays.

In the framing of floors, some persons leave the stair trimmer out until the stairs are put up, and then the trimmer is put up by the stair case hand, or joiner.

PLATE X.

Fig. 1 section of a double floor, with a girder, taken trans versely to the bridging joists

A section of girder.

B C, B C binding joists.

d, d, d, &c. ends of bridging joists.

e, e, e, &c. ends of ceiling joists, chace mortised into binding joists.

Fig. 2 section of a double floor, taken transversely to the binding joist.

A, A sections of the binding joists.

B C part of a bridging joist.

D É ceiling joists.

E F, E F parts of ceiling joists.

Figures 3, 4, 5, 6, show the manner of **scarfing** or lengthening of beams.

Fig. 3 an oblique plain scarf.

Fig. 4 a single oblique tabled scarf.

Fig. 5 a parallel scarf keyed together.

Fig. 6 the method of building beams with small pieces.

The third, fourth, and fifth figures must be firmly bolted with at least two bolts. Fig. 4 and 5 have each an opening for a key to be driven through, which must be done previously to the bolting. These beams would be much stronger at the scarfing, if an iron strap were placed on each side of it, in order to resist the heads and nuts of the screws more effectually than the wood.

Fig. 7 a truss for a span roof.

A, A wall plates.

B C tie beam.

C D king post, crown post, or middle post.

E F, E F struts.

g h, g h puncheons.

I G, I G principal rafters.

K, K pole plate.

Carpentry Plate X.

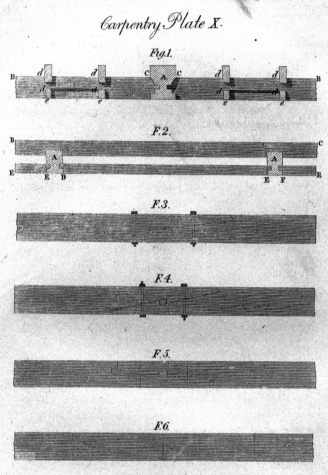

Fig.1.

F.2.

F.3.

F.4.

F.5.

F.6.

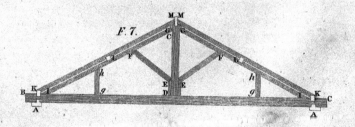

F.7.

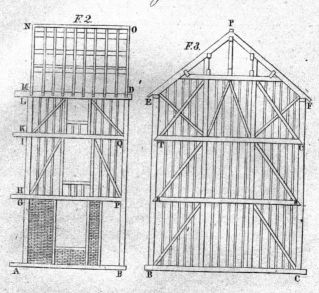

Carpentry Plate XI.

F.2.

F.3.

F.4.

F.5.

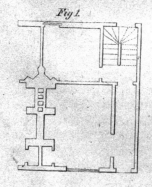

Fig 1.

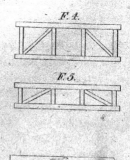

L, L sections of purlines.

K M, K M small rafters.

M M ridge piece section

PLATE XI

The framing for a small wooden house, the lower story constructed of 9 inch brick work, being more secure against external violence, and the upper part of 4 and a half inch stud work, to be covered with lath and plaster. This house is supposed to be constructed where timber is abundant, and brick or stone expensive. The ground story, Fig. 1, consists of a passage, front and back parlour; the one-pair story may be a drawing room, and back room, which may communicate by means of a pair of folding doors; the upper story, which is partly taken out of the roof, may be divided into bed rooms. If two adjoining houses were to be built on the present plan, placing the fire places of the contiguous back to back, so that the same wall, containing the flues, may be common to both, it would not only be a great saving, but strengthen the whole. The partition between the back rooms of the two houses is of wood, and the fire place is placed in the angle of each room, the brick work being continued from the front in order to receive it. The end or gable, is constructed entirely of stud work, to be lathed and plastered. Not only two contiguous houses may be done in this manner, but any series of houses forming a street, by constructing every alternate wall with flues, and every other intervening wall of stud work. The rear fronts will consist entirely of stud work. Wooden houses ought always to stand upon a stone or brick foundation; if, instead of the parlour, the front room were a shop, and the window extending from the door to the wall, then there would be no occasion for any brick work, and the whole would be constructed of stud work, excepting the party wall for the flues. Houses constructed of wood are forbidden in London, by the Building Act; also all interior timbers, within a certain

Nos. 5 & 6.   H

distance of chimnies, as the foregoing abstract which contains what belongs to the carpenter, shows : however, they are much used in country towns, where they are not bound under such restrictions.

Fig. 1 plan.

Fig. 2 elevation.

Fig. 3 gable flank, or division between houses.

A B, B C ground plates, or ground sills.

B D, B E, C F principal posts, extending the whole height of the building, from the ground plate to the roof plate.

A G, H I, K L story posts : all intermediate posts are also called story posts, which extend in altitude from floor to floor.

G P, I Q, R S, T U bressummers, supported by the story posts : the bressummers R S, T U are also interstices, being framed between posts, which in this example are principal posts.

M N, D O Fig. 2 the edges, E P, P F the sides of the extreme rafters.

All the oblique pieces, or those which are placed diagonally within the framing, are called braces.

The tie beam is not placed at the feet of the rafters, but higher, in order to give head room, in consequence of which a brace is extended from the foot of each story post, adjacent to the middle, in the upper story, to each rafter foot ; and as these braces perform the office of ties in this situation, they ought to be well strapped at the ends.

Fig. 4 a longitudinal purline truss.

Fig. 5 a longitudinal truss, placed vertically under the ridge for supporting the intermediate rafters, and restraining them from descending down the inclined plane, and thereby preventing all lateral pressure from the walls ; for it is evident, that if the upper ends of the rafters are held in their situation, the lower ends would describe vertical circles, and from their gravity would descend, and consequently approach nearer together, and therefore, instead of pushing out the walls, would rather have a tendency to draw them in. This principle, as well as trussing the inclined sides of

a roof, was discovered by the author many years ago, in consequence of a dispute, in which he was chosen arbiter on behalf of
the architect; but the principle was so bad, that he was under the
disagreeable necessity of giving judgment in favour of the contractor.

LAW REGULATING BUILDINGS IN THE CITY OF NEW YORK.

*The fire limits include all that part of the city laying south of a line
beginning upon the North River, opposite Spring street, then running up Spring street to Broadway, up Broadway to Art street,
from Broadway along Houston street to the Bowery, down the
Bowery to Division street, from thence up Division street to Governeur streets, down Governeur street to the East River, including
one hundred feet on the northerly and easterly sides of said line,
except that of Division street; and the regulations of buildings
extend to one hundred feet north of Fourteenth street.*

*An Act to amend the Acts heretofore passed for the prevention of
Fires in the City of New York.—Passed April 20th, 1830.*

*The People of the State of New York, represented in Senate and
Assembly: Do enact as follows—*

Walls—Roof.

§ 1. The outside and party walls of all dwelling houses, store
houses, and other buildings hereafter to be erected, or built within
the fire limits of the City of New York, as the same now exist, or
may hereafter be extended, shall be constructed of stone or brick.

2. The outside and party walls of such buildings shall not be
less than eight inches thick, except flues of chimnies, in any part

thereof; and the party or end walls of such buildings shall rise and be extended to the roof, and so far through the same as to meet and be joined to the slate, tile or other covering thereof, by a layer of mortar or cement.

3. The planking or sheeting of the roof of any such building shall in no case be extended across the party or end walls thereof; and all such buildings, and the top and sides of all dormer windows therein, shall be roofed or covered with tile, slate or other fire proof material.

4. All beams or other timbers in the party walls of such buildings shall be separated from each other at least four inches, by brick or mortar; and all plate pieces in the front or rear walls thereof shall recede from the outside of the wall at least four inches; and such wall shall be built up and extended to the slate or other fire-proof covering of the roof.

5. All discharging or arch pieces, used in the chimneys of any such building, shall recede from any flue in any such chimney at least four inches. No such chimney shall be started or built upon the floor of the building, or be cut off to be supported below by wood; and all hearths shall be supported with arches of stone or brick.

6. No timber shall be used in the front or rear of any building, within such fire limits, where stone is now commonly used. Each lintel on the inside of the front or rear wall of every such building shall have a secure brick arch over it, and no bond timber in any wall thereof shall, in width and thickness, exceed the width and thickness of a course of brick; and such bond timber shall be laid at least eighteen inches apart from each other on either side of any wall respectively.

7. All wooden gutters of any such building over thirty feet in height from the level of the side-walk to the foot of the rafters, shall be lined or covered on the upper surface thereof with copper, zinc, or other fire proof material.

8. All scuttles on any such buildings shall be made or covered with copper, zinc, iron, or other fire proof material; and all win-

dow shutters and doors in the rear of any such building, if such building be over thirty feet in height as aforesaid, which shall be used as a warehouse, or storehouse for goods, shall be made of iron or copper.

9. All plate pieces in any such building as is described or mentioned in the first section of this Act, shall be firmly secured with iron anchors, and the cornice of every such building shall be hung in iron anchors.

10. The anchors so to be used at each end of any such cornice shall be at least four feet long, including an angle of at least one foot, and shall be worked or built into the side or end walls of the building ; and such anchors used for supporting the centre of the cornice shall return down the front of the building on the inner side, and shall be firmly secured to the front beam.

11. Every building of more than thirty feet in height from the level of the sidewalk to the foot of the rafters, which shall hereafter be erected or built to the southward of a line distant one hundred feet north of the northerly side of Fourteenth street, shall be subject to all the provisions of this Act.

12. Every building within the fire limits, as the same now exist, or may hereafter be extended, which may hereafter be damaged by fire to an amount equal to two-thirds of the whole value thereof, after the lapse of at least fifteen years from the time of its first erection, shall be repaired or rebuilt according to the provisions of this Act.

13. The amount or extent of such damage may be determined by three indifferent persons residing in the said City, one of whom shall be appointed by the owner or owners of the building, another by the fire-wardens of the ward in which such building is situated, and the third by the two persons so appointed, and the decision in writing of such three persons, or of any two of them, shall be final and conclusive, in all cases where such mode of determining the extent of such damage shall have been agreed upon.

14. All roofs, steeples, cupolas, and spires of churches, or other public buildings, (where such public building shall stand at least

G 2

ten feet distant from any and every other building,) may be covered with boards or shingles.

15. Public buildings, as mentioned in the last preceding section,
are hereby defined to be such buildings as shall be owned and occupied for public purposes, by this State, the United States, the Corporation of the City of New York, or the Public School Society.

16. All privies not exceeding ten feet square and fifteen feet in
height, and all fire engine houses belonging to the Corporation of
the said City, and all lime and ferry houses which shall be erected
with the express permission of the said Corporation, may be built
and covered with wood, boards, or shingles.

17. The owner or owners of any building who shall violate any
of the foregoing provisions of this Act, shall, for every such offence,
forfeit and pay the sum of five hundred dollars ; and every builder
who shall be employed, or assist in so doing, whether he be an
owner of such building or not, shall, for every such offence, forfeit
and pay the additional sum of two hundred and fifty dollars.

18. The foregoing provisions of this Act shall not apply to any
building heretofore erected by any lessee, or lessees, or other person possessed of a leasehold interest in any lands, tenements, or
hereditaments, and which by any express exception in any law
heretofore passed, relative to the prevention of fires in the City of
New York, would be exempt from the provisions of such law.

19. All ash holes or ash houses within the said City shall be
built of stone or brick, without the use of wood in any part thereof.

20. No wooden shed exceeding twelve feet in height, at the
peak or highest part thereof, shall be erected within the fire limits
of the said City, as the same now exist, or may hereafter be extended.

21. No wooden building shall be raised, enlarged, or built upon,
or removed from one lot to any other lot, within such fire limits as
the same now exist, or may hereafter be extended.

22. The owner or owners of any ash house, or ash hole, wooden
shed, or wooden building, who shall violate any of the provisions
of the nineteenth, twentieth, or twenty-first sections of this Act,

and every master builder who may be employed, or assist therein, shall, for every such offence, severally forfeit and pay the sum of two hundred and fifty dollars: and such owner or owners shall forfeit and pay the additional sum of fifty dollars for every twenty-four hours during which such ash house or ash hole, wooden shed, or wooden building shall remain, in violation of any such provision after due notice shall have been given to remove the same.

23. Every house, shed, or other building of any description whatsoever hereinbefore mentioned, which shall hereafter be erected, built, roofed, repaired, altered, enlarged, built upon, or removed, contrary to any of the foregoing provisions of this Act, shall be deemed a common nuisance.

24. It shall not be lawful for any person or persons to have or keep any quantity of gunpowder exceeding twenty-eight pounds in weight, in any one house, store, building, or other place in the City of New York, to the southward of a line running through the centre of Fourteenth street, from the North to the East River, or to lade, receive, have or keep any greater quantity of gunpowder than as aforesaid, on board of any ship, vessel, boat, or other water craft whatever, within three hundred yards from any wharf, pier or slip in that part of the City lying southward of the said line.

25. All gunpowder which may be kept in the said City, or on board of any ship, vessel, boat, or other water craft, to the southward of the line mentioned in the last section, shall be kept in stone jugs or tin canisters, which shall not contain more than seven pounds each.

26. If any person or persons shall have or keep any gunpowder in the City of New York, or on board of any ship, vessel, boat, or other water craft, to the southward of the said line, in any manner contrary to the foregoing provisions of this Act, either as to quantity or as to the manner of keeping the same, he, she, or they shall forfeit and pay the sum of one hundred and twenty-five dollars for every hundred pounds of gunpowder so had or kept, and in that proportion for a greater or less quantity; and all such gunpowder shall be forfeited to the Fire Department of the said City.

27. The commander, or owner or owners of every ship, or other vessel, arriving in the harbour of New York, and having more than twenty-eight pounds of gunpowder on board, shall within forty-eight hours after such arrival, and before such ship or vessel shall approach within three hundred yards of any wharf, pier, or slip, to the southward of a line drawn through the centre of Fourteenth-street as aforesaid, cause the said gunpowder to be landed by means of a boat, or boats, or other small craft, at any place without the said limits, which may be most contiguous to any magazine for storing gunpowder, and shall cause the said gunpowder to be stored in such magazine, on pain of forfeiting the same to the Fire Department of the City of New-York.

28. It shall be lawful either to proceed with any such ship or vessel to sea, within forty-eight hours after her arrival, or to transship such gunpowder from one ship or vessel to another, for the purpose of immediate exportation, without landing such gunpowder as in the last section is directed; but in neither case shall it be lawful to keep such gunpowder for a longer time than forty-eight hours in the harbour of New-York, or to approach with the same within three hundred yards of any wharf, pier, or slip in the said City, to the southward of the line specified in the last section, on pain of forfeiture as therein mentioned.

29. All gunpowder which shall be conveyed or carried through any of the streets of the City of New-York, in any cart, carriage, wagon, wheelbarrow, or otherwise, shall be secured in tight cask or kegs, well headed and hooped, each of which shall be put into and entirely covered with a leather bag or case, sufficient to prevent any of such gunpowder from being spilled or scattered; and all gunpowder which shall be conveyed or carried through any of the said streets, in any other manner than as above directed, shall be forfeited to the Fire Department of the said City.

30. In every case of a violation of any provision of this Act, where the penalty prescribed thereby for such violation is the forfeiture of any gunpowder to the said Fire Department, it shall be lawful for any fire warden of the said City to seize such

gunpowder in the day time, and to cause the same to be conveyed to any magazine used for the purpose of storing gunpowder.

31. It shall be the duty of every person who shall have made any such seizure forthwith to inform the Mayor or Recorder and any two Aldermen of the said City thereof; and the said Mayor or Recorder and Aldermen shall thereupon inquire into the facts and circumstances of such alleged violation and seizure, for which purpose they may summon any person or persons to testify before them, and they shall have power, in their discretion, to order any gunpowder so seized to be restored.

32. Whenever any inhabitants of the said City shall make oath before the Mayor, or Recorder, or any two Aldermen, or any two of the Special Justices thereof, of any fact or circumstance which, in the opinion of the said Mayor, Recorder, Aldermen, or Special Justices, shall afford a reasonable cause of suspicion, that any gunpowder has been brought, or is kept, within the said City, or in the harbour thereof, contrary to any provision contained in this Act, it shall be lawful for the said Mayor, Recorder, Aldermen, or Special Justices, to issue his or their warrant or warrants, under his or their hand and seal, to any sheriff, marshal, constable, or other fit person or persons, commanding him or them to search for such gunpowder in the day time, wheresoever the same may be in violation of this Act, and to seize and take possession of the same if found; but no person having or acting under any such search warrant shall take advantage thereof to serve any civil process whatsoever.

33. It shall be lawful for any person or persons who by virtue of any such warrant, shall have seized any gunpowder, to cause the same within twelve hours, in the day time, after such seizure, to be conveyed to any magazine used for storing gunpowder, and unless the said Mayor or Recorder and any two Aldermen of the said City, should in the manner directed by the thirty-first section of this Act, order the same to be restored, such gunpowder shall be detained in such magazine until it shall be determined by due

I

course of law, whether the same may have become forfeited by virtue of this Act.

34. All actions or suits for the recovery of any gunpowder which may have been seized and stored in any magazine, by virtue of this Act, or for the value thereof, or for damages sustained by the seizure or detention thereof, shall be brought against the Fire Department of the City of New York, and shall be commenced within three calendar months next after such seizure shall have been actually made; and in case no such action or suit shall have been commenced within such period, such gunpowder shall be deemed absolutely forfeited to the said Fire Department, and may be immediately delivered to the proper officers thereof for its use. No penal damages shall be recovered in any such action or suit, and such gunpowder may at any time during the pendency of any such action or suit, by consent of the parties thereto, be removed from any magazine where the same may have been stored, or may be sold, and the moneys arising from such sale may be paid into the Court where such suit or action may be pending, to abide the event thereof.

35. Nothing contained in this Act shall be construed to apply to any ship or vessel of war in the service of the United States, or of any foreign government while lying distant three hundred yards or upwards from the wharves, piers or slips of the said City.

36. If any gunpowder exceeding twenty-eight pounds in quantity, shall be found in the possession or custody of any person, by any fireman of the said City, during any fire or alarm of fire therein, it shall be lawful for such fireman to seize the same without any warrant, and to report such seizure without delay to the Mayor or Recorder of the said City, and it shall be determined by the said Mayor or Recorder and any two Aldermen of the said City, in the manner directed by the thirty-first section of this Act whether such gunpowder should be restored, or the same shall be conveyed to a magazine for storing gunpowder and there detained, until it be decided by due course of law, whether such gunpowder be forfeited by virtue of this Act.

37. No greater quantity of sulphur than ten hundred weight, or of hemp or flax, than twenty hundred weight, or of pitch, tar, turpentine, rosin, spirits of turpentine, varnish, linseed oil, oil of vitriol, aquafortis, æther, or shingles, than shall be allowed by the Common Council of the City of New York, shall be put, kept, or stored in any one place in the said City, to the southward of a line drawn through the centre of Fourteenth street, unless with the permission of the said Common Council.

38. Every person who shall violate either of the provisions of the last section, shall for every such offence forfeit and pay the sum of twenty-five dollars; and in case any such person or persons shall neglect or refuse to remove any of the articles prohibited by the said section, within such time as may be allowed for that purpose by the Mayor or Recorder, or any two Aldermen of the said City, he, she or they shall, for every such neglect or refusal, forfeit and pay an additional sum of twenty-five dollars.

39. Nothing hereinbefore contained shall be construed to prohibit any ship chandler from keeping at any time, in any enclosure in the said City, any quantity of pitch, tar, rosin, or turpentine, not exceeding twenty barrels in the whole.

40. All pecuniary penalties imposed by this Act, may be sued for and recovered with costs of suit, in any court having cognizance thereof, by the proper officers of the Fire Department of the said City, for the use of the said Fire Department.

41. All actions for any forfeiture or penalty incurred under this Act, shall be commenced within one year next after the time of incurring such forfeiture or penalty.

42. All laws or parts of laws, heretofore passed, inconsistent with the provisions of this Act, are hereby declared to be repealed; but such repeal shall not affect any suit or prosecution already commenced, or any penalty, forfeiture, or offence already incurred or committed under any such law or part of a law.

LIEN LAW.

An Act for the better security of Mechanics and others, erecting Buildings in the City and County of New York.—Passed April 20, 1830.

The People of the State of New York, represented in Senate and Assembly: Do enact as follows—

§ 1. Every mechanic, workman, or other person, doing or performing any work towards the erection, construction, or finishing of any building in the city of New York, erected under a contract in writing, between the owner and builder, or other person, whether such work shall be performed as journeyman, labourer, cartman, sub-contractor, or otherwise, and whose demands for work and labour done and performed towards the erection of such building, has not been paid and satisfied, may deliver to the owner of such building an attested account of the amount and value of the work and labour thus performed and remaining unpaid, and thereupon such owner shall retain, out of his subsequent payments to the contractor, the amount of such work and labour for the benefit of the person so performing the same.

2. Whenever any account of labour performed on a building erected under a contract in writing, as aforesaid, shall be placed in the hands of the owner of such building, or his authorized agent, it shall be the duty of such owner or agent to furnish his contractor with a copy of such papers, in order that if there shall be any disagreement between such contractor and his creditor, they may, by amicable adjustment between themselves, or by arbitration, ascertain the true sum due, and if the contractor shall not, within ten days after the receipt of such papers, give the owner written notice that he intends to dispute the claim, or if in ten days after giving such notice, he shall refuse or neglect to have the matter adjusted as aforesaid, he shall be considered as assenting to the demand, and the owner shall pay the same when it becomes due.

3. If any such contractor shall dispute the claim of his journeyman or other person for work and labour performed as aforesaid, and if the matter cannot be adjusted amicably between themselves, it shall be submitted, on the agreement of the parties, to the arbitrament of three disinterested persons, one to be chosen by each of the parties, and one by the two thus chosen, and the decision in writing, of such three persons, or any two of them, shall be final and conclusive in the case submitted.

4. Whenever the amount due shall be adjusted and ascertained, as above provided, and if the contractor shall not, within ten days after it is so adjusted and ascertained, pay the sum due to his creditor, with the costs incurred, the owner shall pay the same out of the funds as above provided, and which amount due may be recovered from the said owner by the creditor of the said contractor, in an action for money had and received to the use of said creditor, and to the extent in value of any balance due by the owner to his contractor under the contract with him at the time of the notice first given as aforesaid, or subsequently accruing to such contractor under the same, if such amount shall be less than the sum due from the said contractor to his creditor.

5. If by collusion, or otherwise, the owner of any building erected by contract in writing, as aforesaid, shall pay to his contractor any money in advance of the sum due on said contract, and if the amount still due the contractor, after such payment has been made, shall be insufficient to satisfy the demand, made in conformity with the provisions of this act, for work and labour done and performed, the owner shall be liable to the amount that would have been due at the time of his receiving the account of such work, in the same manner as if no such payment had been made.

INDEX

EXPLANATION OF TERMS

USED IN

CARPENTRY.

N. B. *This Mark* § *refers to the preceding Sections, according to the Number.*

A.

B.

BACK OF A HIP is the upper edge of a rafter, between the two sides of a hipped roof formed to an angle so as to range with the rafters on each side of it.

BAULK, a piece of foreign fir, or deal, being the trunk of a tree of that species of wood, generally brought to a square, for the use of building. In London the term is only applied to small lengths, from 18 to 25 feet, generally under 10 inches thick, having a considerable taper, and the wains left, so that the baulk is not brought to a square. In some parts of the country these obtain the name of Dram timber, as coming from the place of that name. In London the largest pieces of timber, such as Memel,

Dantzic, &c. seem to have no common appellation, being familiarly called pieces of timber, and frequently by the vulgar name of sticks; these expressions seem to define nothing, as they apply equally to all sizes. Different names seem to obtain in different parts of the country : in some parts of the north, large pieces of fir wood are called logs; but in London log is restricted to the largest pieces of oak or mahogany.

BEAM, a horizontal timber, used to resist a force, or weight, as a tie-beam, where it acts as a string, or chain, by its tension; as a collar beam, where it acts by compression; as a bressummer, where it resists a transverse insisting weight.

BEARER, any thing used by way of support to another.

BEARING, the distance that a beam or rafter is suspended in the clear : thus if a piece of timber rests upon two opposite walls, the span of the void is called the bearing, and not the whole length of the timber.

BEETLE, § 17.

BOARD, a substance of wood contained between two parallel planes; as when the baulk is divided into several pieces by the pit saw, the pieces are called boards. The section of boards is sometimes, however, of a triangular, or rather a trapazoidal form, that is with one edge very thin : these are called feather edged boards.

BOND TIMBER, § 33.

BRACE, a piece of slanting timber, used in truss partitions, or in framed roofs, in order to form a triangle, and thereby rendering the frame immovable; when a brace is used by way of support to a rafter, it is called a strut. Braces in partitions, and span roofs, are always, or should be, disposed in pairs, and placed in opposite directions.

BREAKING DOWN, in sawing, is dividing the baulk into boards or planks; but if planks are sawed longitudinally through their thickness, the saw-way is called a ripping cut, and the former a breaking cut.

BRESSUMMER, or BREASTSUMMER, a beam supporting a superincumbent part of an exterior wall, and running longitudinally below that part. *See Summer.*

Bridging Joists are the smallest beams in naked flooring, for supporting the boarding for walking upon. *See Plate.*

Bring up. *See Carry-up.*

C.

Camber is the convexity of a beam upon the upper edge, in order
to prevent its becoming straight or concave by its own weight,
or by the burden it may have to sustain, in course of time.

Camber Beams are those used in the flats of truncated roofs, and
raised in the middle with an obtuse angle, for discharging the
rain water towards both sides of the roof.

Cantilevers are horizontal rows of timbers, projecting at right
angles from the naked part of a wall, for sustaining the eaves or
other mouldings. Sometimes they are planed on the horizontal and vertical sides, and sometimes the carpentry is rough and
cased with joinery.

Carcass of a Building, is the naked walls, and the rough timber
work of the flooring and quarter partitions, before the building
is plastered, or the floors laid.

Carpenter's Square, § 21.

Carpentry, § 1.

Carry-up, a term used in discourse among builders and workmen,
denoting that the walls, or other parts, are intended to be built
to a certain given height; as the carpenter will say to the bricklayer, Carry-up that wall; carry-up that stack of chimnies, *i. e.*
build up that wall or stack of chimnies.

Chisels, § 6, 7, and 8.

Crown Post, the middle post of a trussed roof. *See King Post.*

D.

Deal Timber, the timber of the fir tree, as cut into boards, planks,
&c. for the use of building.

Discharge, is a post trimmed up under a beam, or part of a build
ing which is weak, or overcharged by weight.

Dormer, or Dormer Window, is a projecting window in the roof of a house, the glass frame, or casement, being set vertically, and not in the inclined sides of the roof; thus dormers are distinguished from sky-lights, which have their sides inclined to the horizon.

DOVETAIL NOTCH, § 27.

DRAGON BEAM, the piece of timber which supports the hip raiter, and bisects the angle formed by the wall plates,

DRAW BORE PINS. *See Joinery.*

E.

ENTER, when the end of a tenon is put into a mortise, it is said to enter the mortise.

ENTERTICE. *See Intertie.*

F.

FEATHER-EDGED BOARDS. *See Board.*

FILLING-IN-PIECES, short timbers less than the full length, as the jack rafters of a roof, the puncheons, or short quarters in partitions, between braces and sills, or head-pieces.

FIR POLE, small trunks of fir trees, from 10 to 16 feet in length, used in rustic buildings, and out-houses.

FIRMER CHISEL, § 7.

FLOOR. *See Naked Flooring.*

FOUNDATIONS, § 33.

FURRINGS, are slips of timber nailed to joists or rafters, in order to bring them to a level, and to range them into a straight surface, when the timbers are sagged, either by casting or by a set, which they have obtained by their weight in length of time.

G.

GAIN, a term now out of use. *See Tusk.*

GAUGE, § 11.

GIMLET, § 9.

GIRDER, the principal beam in a floor for supporting the binding joists.

Nos. 6.   A

GROOVED NOTCH, § 29. *See Plate* II.

GROUND PLATE, or SILL, is the lowest plate of a wooden building for supporting the principal and other posts. *See Plate* V.

H.

HAMMER, § 15.

HAND SAW, § 3.

HOOK PINS, § 20.

HANDSPIKE, a lever for carrying a beam, or other body, the weight being placed in the middle, and supported at each end by a man.

I.

INTERTIE, a horizontal piece of timber, framed between two posts in order to tie them together.

JACK TIMBER, a timber shorter than the whole length of other pieces in the same range.

JACK RAFTERS are all those short rafters which meet the hips.

JACK RIBS are those short ribs which meet the angle ribs, as in groins, domes, &c.

JOGGLE PIECE is a truss post, with shoulders and sockets for abutting and fixing the lower ends of the struts.

JOINING OF TIMBERS, §§ 22, 23, 24, 25, 26, 27.

JOISTS are those beams in a floor which support, or are necessary in the supporting of the boarding or ceiling, as the binding, bridging, and ceiling joists; girders are, however, to be excepted, as not being joists.

JUFFERS, stuff of about four or five inches square, and of several lengths. This term is out of use, though frequently found in old books.

K.

KING POST, the middle post of a stuffed roof, for supporting the tie-beam at the middle, and the lower ends of the struts.

KERF, the way made by the saw in sawing timber.

L.

LAW regulating buildings in the City of New York, page 67.

LEVEL, an instrument used for levelling floors, § 12.

LIEN LAW, page 76.

LINTELS, short beams over the heads of doors and windows, for supporting the inside of an exterior wall, or the superincumbent part over doors, in brick or stone partitions.

LUTHORN windows. *See Dormer.*

M.

MALLET, § 16.

MORTISE and TENON, § 31.

N.

NAKED FLOORING, the timber work of a floor for supporting the boarding or ceiling, or both.

NOTCHING, § 28, 29.

P.

PITCH OF A ROOF, the inclination which the sloping sides make with the plane or level of the wall-plate; or it is the proportion which arises by dividing the span by the height. Thus if it is asked, What is the pitch of such a roof? the answer is, quarter, 3 quarters, or half; when the pitch is half, the roof is a square, which is the highest now in use, or that is necessary in practice.

PLANK, all boards above nine inches wide, are called planks.

PLATE, a horizontal piece of timber in a wall, generally flush with the inside, for resting the ends of beams, joists, or rafters, and is therefore denominated floor, or roof plates, accordingly.

PLUMB RULE, § 14.

POSTS, all upright or vertical pieces of timber, whatever, as truss posts, door posts, quarters in partitions, &c.

PRICK POSTS, intermediate posts in a wooden building framed between principal posts.

PRINCIPAL POSTS, the corner posts of a wooden building. *See Plate V.*

PUDLAIES, pieces of timber to do the office of handspikes.

PUNCHEONS, any short posts of timber; the small quarterings in a stud partition above the head of a door, are called puncheons.

PURLINES, the horizontal timbers in the sides of a roof, for supporting the spars or small rafters.

Q.

QUARTERS, the timbers to be used in stud partitions, bond in walls, &c.

QUARTERING, the stud work of a partition.

R.

RAFTERS, all the inclined timbers in the sides of a roof, as principal rafters, hip rafters, and common rafters, which are otherwise called, in most countries, spars.

RAISING PLATES, or TOP PLATES, are the plates on which the roof is raised.

REBATED NOTCH, § 28.

RIDGE, the meeting of the rafters on the vertical angle of the roof. *See Plate V.*

RIPPING CHISEL, § 3.

RIPPING SAW, § 3.

ROOF, the covering of a house; but the word is used in carpentry for the wood work which supports the slating and other covering.

S.

SAW, § 3.

SHAKEN STUFF. Such timber as is rent or split by the heat of the sun, or by the fall of the tree, is said to be shaken.

SHINGLES, thin pieces of wood used for covering, instead of tiles, &c.

SHREADINGS, a term not much used at present. *See Furrings.*

SKIRTS of a Roof, the projecture of the eaves.

SLEEPERS, pieces of timbers for resting the ground joists of a floor upon, or for fixing the planking to in a bad foundation. The term was formerly applied to the valley rafters of a roof.

SOCKET CHISEL, § 6.

SPARS, the term by which the common rafters of a roof are best known in almost every provincial town in Great Britain, though generally called in London common rafters, in order to distinguish them from the principal rafters.

STANCHEONS. *See Puncheons.*

STRUTS, pieces of timber which support the rafters, and which are supported by the truss posts.

SUMMER, a large beam in a building, either disposed in an outside wall, or in the middle of an apartment, parallel to such wall. When a summer is placed under a superincumbent part of an outside wall, it is called a bressummer, as it comes in a breast with the front of the building.

STUDWORK, § 33.

T.

TEMPLETS, § 33.

TENON, § 30.

TIE, a piece of timber placed in any position acting as a string or tie, to keep two things together which have a tendency to a more remote distance from each other.

TIMBERS, how joined, §§ 22, 23, 24, 25, 26, 27.

TRIMMERS are joists into which other joists are framed.

TRIMMING JOISTS, the two joists into which a trimmer is framed.

TRUNCATED ROOF, is a roof with a flat on the top.

TRUSS, a frame constructed of several pieces of timber, and divided into two or more triangles by oblique pieces, in order to pre-

H 2

vent the possibility of its revolving round any of the angles of the frame.

TRUSS POST, any of the posts of a trussed roof, as king post, queen post, or side post, or posts into which the braces are formed in a trussed partition.

TRUSSED ROOF is one so constructed within the exterior triangular frame, so as to support the principal rafters and the tie beam, at certain given points.

TUSK, the beveling upper shoulder of a tenon, in order to give strength to the tenon.

V.

VALLEY RAFTER, that which is disposed to the internal angle of a roof.

W.

WALL PLATES, are the joists' plates, and raising plates.

JOINERY.

§ 1. JOINERY is a branch of Civil Architecture, and consists of the art of framing or joining together wood for internal and external finishings of houses; as the coverings and linings of rough walls, or the coverings of rough timbers, and of the construction of doors, windows, and stairs.

Hence joinery requires much more accurate and nice workmanship than carpentry, which consists only of rough timbers, used in supporting the various parts of an edifice. Joinery is used by way of decoration only, and being always near to the eye, requires that the surfaces should be smooth, and the several junctions of the wood be fitted together with the greatest exactness.

Smoothing of the wood is called planing, and the tools used for the purpose, planes.

The wood used is called stuff, and is previously formed into rectangular prisms by the saw; these prisms are denominated battens, boards, or planks, according to their dimensions in breadth or in thickness. For the convenience of planing, and other operations, a rectangular platform is raised upon four legs, called a bench.

§ 2. *The Bench.* PL. 12. FIG. 12.

Consists of a platform A B C D called the top, supported upon four legs, E, F, G, H. Near to the further or fore end A B is an upright rectangular prismatic pin *a*, made to slide stiffly in a mortise through the top. This pin is called the bench hook, which ought to be so tight as to be moved up or down only by a blow of a hammer or mallet. The use of the bench hook is to keep the stuff steady, while the joiner, in the act of planing, presses it forward against the bench hook. D I a vertical board fixed to the legs, on the side of the bench next to the workman, and made flush with the legs : this is called the side board. At the farther end of the side board, and opposite to it, and to the bench hook, is a rectangular prismatic piece of wood *b b*, of which its two broad surfaces are parallel to the vertical face of the side board : this is made moveable in a horizontal straight surface, by a screw passing through an interior screw fixed to the inside of the side board, and is called the screw check. The screw and screw check are to-gether called the bench screw; and for the sake of perspicuity, we shall denominate the two adjacent vertical surfaces of the screw check, and of the side board, the checks of the bench screw. The use of the bench screw is to fasten boards between the checks, in order to plane their edges ; but as it only holds up one end of a board, the leg H of the bench and the side board are pierced with holes, so as to admit of a pin for holding up the other end, at various heights, as occasion may require. The screw check has also a horizontal piece mortised and fixed fast to it, and made to slide through the side board, for preventing it turning round, and is therefore called the guide.

Benches are of various heights, to accommodate the height of the workman, but the medium is about two feet eight inches. They are ten or twelve feet in length, and about two feet six inches in width. Sometimes the top boards upon the farther side are made only about ten feet long, and that next the workman twelve feet, projecting two feet at the hinder part. In order to keep the bench and work from tottering, the legs, not less than three inches and a half square, should be well braced, particularly the two legs on the working side. The top board next to the workman may be from one and a half to two inches thick : the thicker, the better for the work; the boards to the farther side

may be about an inch, or an inch and a quarter thick. If the workman stands on the working side of the bench, and looks across the
bench, then the end on his right hand is called the hind end, and that
on his left hand the fore end. The bench hook is sometimes covered
with an iron plate, the front edge of which is formed into sharp teeth
for sticking fast into the end of the wood to be planed, in order to
prevent it from slipping; or, instead of a plate, nails are driven
obliquely through the edge, and filed into wedge-formed points. Each
pair of end legs are generally coupled together by two rails dove-
tailed into the legs. Between each pair of coupled legs, the length
of the bench is generally divided into three or four equal parts, and
transverse bearers fixed at the divisions to the side boards, the
upper sides being flush with those of the side boards, for the purpose of supporting the top firmly, and keeping it from bending.
The screw is placed behind the two fore legs, the bench hook
immediately before the bearers of the fore legs, and the guide at
some distance before the bench hook. For the convenience of
putting things out of the way, the rails at the ends are covered
with boards; and for farther accommodation, there is in some
benches a cavity, formed by boarding the under edges of the side
boards before the hind legs, and closing the ends vertically, so that
this cavity is contained between the top and the boarding under
the side boards; the way to it is by an aperture made by sliding
a part of the top board towards the hind end: this deposit is called
a locker.

§ 3. *Joiners' Tools.*

The bench planes are, the jack plane, the fore plane, the trying
plane, the long plane, the jointer, and the smoothing plane; the
cylindric plane, the compass and forkstaff planes; the straight
block, for straightening short edges. Rebating planes are the moving fillister, the sash fillister, the common rebating plane, the side
rebating plane. Grooving planes are the plough and dado grooving

L

planes. Moulding planes are sinking snipebills, side snipebills, beads, hollows and rounds, ovolos and ogees. Boring tools are, gimlets, brad-awls, stock, and bits. Instruments for dividing the wood, are principally the ripping saw, the half ripper, the hand saw, the panel saw, the tenon saw, the carcase saw, the sash saw, the compass saw, the keyhole saw, and turning saw. Tools used for forming the angles of two adjoining surfaces, are squares and bevels. Tools used for drawing parallel lines are guages. Edge tools, are the firmer chisel, the mortise chisel, the socket chisel, the gouge, the hatchet, the adze, the drawing knife. Tools for knocking upon wood and iron are, the mallet and hammer. Implements for sharpening tools are the grinding stone, the rub stone, and the oil or whet stone.

---

§ 4. *Definitions.*

If a plane be set with the under surface upon the wood it is intended to operate upon, and placed before the workman, and if four surfaces are perpendicular to the under surface, each of these surfaces is said to be vertical; the one next the workman is called the hind end, and the opposite one, the fore end, and the two in the direction which the plane works, the sides: the under surface is called the sole, the side of the plane next to the workman is called the right hand side, and the opposite side to that, the left hand side of the plane.

The depth of a plane is the vertical dimension from the top to the under surface; the length of a plane is the horizontal dimension in the direction in which the plane is wrought; the breadth or thickness of a plane is the horizontal dimension at right angles, to the length and depth.

In order to make a distinction between the tool, the under surface is called the sole of the plane.

The reason for being so particular in defining these common

place terms which might be supposed to be known to every one, is from a desire of the author to prevent ambiguity ; as in the term depth, which implies a distance from you in whatever direction it runs, as the depth of a well is the vertical or plumb distance ; but the depth of a house is the distance from the front to the rear wall, and consequently is a horizontal distance

§ 5. *The Jack Plane,* PL. 12. FIG. 1.

Is used in taking off the rough and prominent parts from the surface of the wood, and reducing it nearly to the intended form, in coarse slices, called shavings ; this plane consists of a block of wood called the stock, of about seventeen inches in length, three inches high, and three inches and a half broad. All the sides of the stock are straight surfaces at right angles to each other. Through the solid of the stock, and through two of its opposite surfaces is cut an aperture, in which is inserted a thin metal plate called the iron, one side of the plate consisting of iron, and the other of steel. The side of the opening which joins the iron part, is called the bed, which is a plane surface, making an angle of forty-five degrees with the hind part of the underside of the plane.

The end of the iron next to the bottom is ground to an acute angle off the iron side, so as to bring the steel side to a sharp edge, having a small convexity. The sloping part thus formed, is called the basil of the iron. The iron is fixed by means of a wedge, which is let into two grooves of the same form, on the sides of the opening ; two sides of the wedge are parallel to each other, and to the vertical side of the plane, and consequently to two of the sides of the groove ; the two sides of the grooves, parallel to the vertical sides of the plane are called cheeks, and the two other sides inclined to the bed of the iron are called the abutments or abutment sides : the wedge and the iron being fixed, the opening must be uninterrupted from the sole to the top, and must be no more on the sole side of the plane, than what is sufficient for the thickest shaving to pass with ease ; and as the shaving is discharged at the upper side of the plane, the opening through must expand or increase from the sole to the top, so as to prevent the shavings from sticking. In conformity to analogy, the part of the opening at the sole, which first receives the shaving, is called the mouth. In order for the shaving to pass with still greater ease, the wedge (Pl. 12. Fig. 5.) is forked to cut away in the middle, leaving the prongs to fill the lower parts of the aforesaid grooves. On the upper part of the plane, behind the iron, rises a protuberance, called the tote, so formed to the shape of the hand, and direction of the motion, as to produce the most power in pushing the plane forward.

The bringing of the iron to a sharp cutting edge is called sharpening. The cutting edge of the iron must be formed with a convexity, and regulated by the stuff to be wrought, whether it is hard or soft, cross grained or curling, so that a man may be able to perform the most work, or to reduce the substance most, in a given time. To prevent the iron from tearing the wood to cross grained stuff, a cover is used with a reversed basil, (Pl. 12. Fig. 4.) and fastened by means of a screw, the thin part of which slides in a longitudinal slit in the iron, and the head is taken out by a large hole near the upper end of it. The lower edge of the cover is so formed, as to be concentric or parallel to the cutting-edge of the iron, and fixed at a small distance above it, and to coincide entirely with the steel face. The basil of the cover must be rounded, and not flat, as that of the iron is. The distance between the cutting edge of the iron, and the edge of the cover, depends altogether on the nature of the stuff. If the stuff is free, the edge of the cover may be set at a considerable distance, because the difficulty of pushing the plane forward becomes greater, as the edge of the cover is nearer the edge of the iron, and the contrary when more remote.

The convexity of the edge of the iron depends on the texture of the stuff, whether it is free, cross grained, hard or knotty. If the stuff is free, it is evident that a considerable projection may be allowed, as a thicker shaving may be taken: the extreme edges of the iron must never enter the wood, as this not only retards the progress of working, but chokes and prevents the regular discharge of the shavings at the orifice of the plane.

---

§ 6. *To Grind and Sharpen the Iron.*

When you grind the iron, place your two thumbs under it, and the fingers of both hands above, laying the basil to the stone, and holding it to the angle you intend it shall make with the steel side of it, keeping it steady while the stone is turning, and pressing the iron to the stone with your fingers; and in order to prevent the stone from wearing the edge of the iron into irregularities, move it alternately from edge to edge of the stone with so much pressure on the different parts, as will reduce it to the required convexity; then lift the iron to see that it is ground to your mind: if it is not, the operation must be repeated, and the steel or basil side placed in its former position on the stone, otherwise the basil will be doubled; but if in the proper direction it will be hollow, which will be more as the diameter of the stone is less. The basil being brought to a proper angle, and the edge to a regular curvature, the roughness occasioned by the gritty particles of the grindstone may be taken away, by rubbing on a smooth flat wet stone or Turkey stone, sprinkling sweet oil on the surface; as the basil is generally ground something longer that what the iron would stand, for the quicker despatch of wetting it, you may incline the face of the iron nearer to the perpendicular, rubbing to and fro with the same inclination throughout: having done it to your mind, it may be fixed. When there is occasion to sharpen it again, it is commonly done upon a flat rub stone keeping the proper angle of position as before, then the edge may be finished on the Turkey stone as before: and at every time the iron gets dull or blunt, the sharpening is produced by the rub stone and Turkey stone, but in repeating this often the edge gets so thick that it requires so much time to bring it up, that recourse must be had again to the grindstone.

§ 7. *To Fix and Unfix the Iron.*

In fixing the iron in the plane, the projection of the cutting edge must be just so much beyond the sole of the plane, as the workman may be able to work it freely in the act of planing. This projection is called iron, and the plane is said to have more or less iron as the projection varies: when there is too much iron, knock with the hammer on the fore end of the stock; and the blows will loosen the wedge, and raise the iron in a certain degree, and the head of the wedge must be knocked down to make all tight again: if the iron is not sufficiently raised, proceed again in the same manner, but if too much, the iron must be knocked down gently by hitting the head with a hammer: and thus, by trials, you will give the plane the degree of iron required. When you have occasion to take out the iron to sharpen it, strike the fore end smartly, which will loosen the wedge, and consequently the iron.

---

§ 8. *To Use the Jack Plane.*

In using the jack plane, lay the stuff before you parallel to the sides of the bench, the farther end against the bench hook: then beginning at the hind end of the stuff, by laying the forepart of the plane upon it, lay hold of the tote with the right hand, and pressing with the left upon the fore end, thrust the plane forward in the direction of the fibres of the wood and length of the plane, until you have extended the stroke the whole stretch of your arms; the shaving will be discharged at the orifice: draw back the plane, and repeat the operation in the next adjacent rough part: proceed in this manner until you have taken off the rough parts throughout the whole breadth, then step forward so much as you have planed, and plane off the rough of another length in the same manner: proceed in this way by steps, until the whole length is gone over and rough planed; you may then return and take all the protuberant parts or sudden risings, by similar operations.

§ 9. *The Trying Plane*, Pl. 12. Fig. 2.

Is constructed similar to the jack plane, except the tote of the jack plane is single, and that of the trying plane double, to give greater strength; the length of this plane is about twenty-two inches, the breadth three and a quarter, and the height three and an eighth. Its use is to reduce the ridges made by the jack plane, and to straighten the stuff: for this purpose it is both longer and broader, the edge of the iron is less convex, and set with less projection: but as it takes a broader though finer shaving, it still requires as much force to push it forward.

---

§ 10. *The Use of the Trying Plane.*

The sharpening of the iron, and the operation of planing is much the same as that of the jack plane; when the side of a piece of stuff has been planed first by the jack plane, and afterwards by the trying plane, that side of the stuff is said to be tried up, and the operation is called trying.

When the stuff is required to be very straight, particularly if the broad and narrow side of another piece is to join it, instead of stopping the plane at every arm's length, as with the jack plane, the shaving is taken the whole length, by stepping forwards, then returning, and repeating the operation throughout the breadth, as often as may be found necessary.

---

§ 11. *The Long Plane*

Is used when a piece of stuff is required to be tried up very straight; for this purpose it is both longer and broader than the trying plane, and set with still less iron; the manner of using it is the same. Its length is twenty six inches, its breadth three inches and five eighths, and depth three inches and one eighth.

§ 12. *The Jointer*

Is still longer than the long plane, and is used principally for
planing straight edges, and the edges of boards, so as to make
them join together; this operation is called shooting, and the edge
itself is said to be shot. The length of this plane is about two feet
six inches, the depth three inches and a half, and the breadth three
inches and three fourths. The shaving is taken the whole length
in finishing the joint, or narrow surface.

---

§ 13. *The Smoothing Plane,* PL. 12. FIG. 3.

Is the last plane used in giving the utmost degree of smoothness
to the surface of the wood : it is chiefly used in cleaning off
finished work. The construction of this plane is the same with
regard to the iron, wedge and opening for discharging the shaving,
but is much smaller in size, being in length seven inches and a
half, in breadth three, and in depth two and three quarters, and
differs in form, on account of its having convex sides, and no tote.

There is also this difference in giving the iron a finer set, that
you may strike the hind end instead of the fore part.

---

§ 14. *Bench Planes.*

The jack plane, the trying plane, the long plane, the jointer and
the smoothing plane, are denominated bench planes.

---

§ 15. *The Compass Plane*

Is similar to the smoothing plane in size and shape, but the sole
is convex, and the convexity is in the direction of the length of
the plane. The use of the compass plane is to form a concave

cylindrical surface, when the wood to be wrought upon is bent with the fibres in the direction of the curve, which is in a plane surface perpendicular to the axis of the cylinder. Consequently compass planes must be of various sizes, in order to accommodate different diameters.

---

§ 16. *The Forkstaff Plane*

Is similar to the smoothing plane in every respect of size and shape, except that the sole is part of a concave cylindric surface, having the axis parallel to the length of the plane. The use of the forkstaff plane is to form cylindric surfaces, by planing parallel to the axis of the cylinder. Planes of this description must likewise be of various sizes, to form the surface to various radii : these two last planes are more used by coach-makers than by joiners.

---

§ 17. *The Straight Block*

Is used for shooting short joints and mitres, instead of the jointer, which in such cases would be rather unhandy ; this plane is also made without the tote, and as it is frequently used in straightening the ends of pieces of wood perpendicularly to the direction of the fibres, the iron is inclined more to the sole of the plane, that is, it forms a more acute angle with it : in order that it may cut clean, the inclination of the basil, and the face of the iron, is therefore less on this account : the length of the straight block is twelve inches, its breadth three and one eighth, and depth two and three quarters.

REBATE PLANES IN GENERAL.

§ 18. *The Rebate Plane*

Is used after a piece of stuff has been previously tried on one

Nos. 7 & 8.   M

side and shot on the other, or tried on both sides, in taking away
a part next to one of the arises of a rectangular or oblong section,
the whole part therefore taken away is a square prism, and the
superfices formed after taken away　the prism is two straight
surfaces, forming an internal right angle with each other; so that
the stuff will now have one internal angle and two external angles.
The operation of this reducing the stuff is called rebating. Rebating is either used by way of ornament, as in the sinking of
cornices, the sunk facias of architraves, or in forming a recess for
the reception of another board, so that the edge of this board may
coincide with that side of the rebate, next to the edge of the rebated piece. The length of rebating planes is about nine inches
and a half, the vertical dimension or depth is about three and a
half, they are of various thickness, from one and three quarters to
half an inch. Rebate planes are of several kinds, some have the
cutting edge of the iron upon the bottom, and some upon the side
of the plane. Of these which have the cutting edge on the
bottom, some are used for sinking, and some for smoothing or
cleaning the bottom of the rebate; and these which have the cutting edge upon one side are called side rebating planes, and are
used after the former in cleaning the vertical side of the rebate.
Rebate planes differ from the bench planes, before mentioned, in
their having no tote; the cavity is not open to the top, but the
wedge is made to fit completely, and the shaving is discharged on
one side or other, according to the use of the plane.

------

§ 19. *Sinking Rebating Planes*

Are of two denominations, the moving fillister and sash fillister:
the moving fillister is for sinking the edge of the stuff next to you,
and the sash fillister the farther edge ; consequently these planes
have their cutting edges on the under side.

§ 20. *Of the moving Fillister*, Pl. 13. Fig.

Upon the bottom of the moving fillister is a slip of wood, so regulated by two screws as one of the vertical sides of the slip may be fixed parallel to the edge of the sole ; then the breadth between this side of the slip and the edge of the sole of the plane is equal to the breadth of the rebate. This slip is called a fence, and the vertical side of it next to the stock, the guide ; as the rebate is made upon the right edge of the stuff, the fence is always upon the left side of the sole. The iron between the guide and the right hand edge of the sole of the plane must project the whole breadth of the uncovered part of the sole, otherwise the plane will not sink, so long as it is kept in one position ; the right hand point of the cutting edge of the iron must stand a small degree without the vertical right hand side of the plane ; for if this point of the iron stood within, the situation of the point would also prevent the sinking of the rebate ; it is also necessary that the cutting edge of the iron should stand equally prominent in all parts out of the sole, otherwise the plane cannot make shavings of an equal thickness, and consequently instead of keeping the vertical position, will turn round and incline to the side on which the shavings are thickest, and thus the part cut away will not have a rectangular section, for the bottom of the rebate will not then be parallel to the upper face of the stuff; and the side which ought to have been vertical, will be a kind of ragged curved surface, formed by as many gradations or steps as the depth consists of the number of shavings. Observe, that whatever regulates any plane which takes away a portion of the stuff next to the edge, to cause the part taken away on the upper face of the stuff from the edge to be of one breadth, is called a fence : in like manner, whatever prevents a plane working downwards beyond a certain distance, is called a stop. Therefore the fence regulates the horizontal breadth of what is taken away, and the stop the vertical dimension or depth, and this is to be understood, not only of rebate planes, but of moulding planes, where the moulding is regulated in its horizontal dimension, in the breadth or thickness of the stuff, and the vertical on the adjacent vertical side.

Returning to the moving fillister, the guide is the bottom surface of a piece of metal which is regulated by a screw, so as to move it to the required distance from the sole. Though the bottom of this piece of metal is properly the stop, yet it is altogether called a stop by plane makers and carpenters; but to avoid a confusion of words, we shall call the bottom of the stop the vertical guide. The stop moves in a vertical groove in the side of the fillister, and has a projection with a vertical perforation, which goes farther into the groove, or into the solid of the stock. The stop is placed on the right hand side of the fillister, between the iron and the fore end of the plane, and is moved up and down by a screw, which is inserted in a vertical perforation from the top of the plane to the groove, and passes through the perforation in the projecting part of the stop, which has a female, or concave screw adapted to that cut on the convex screw. The convex screw is always kept stationary by a plate of metal, let in flush with the upper side of the plane; below this plate, and on the same solid with the screw, is a collar, and above, another which projects still farther upwards by way of a lever, for the ease of turning the screw. This part which turns round, is called the thumb screw. It is evident, as the axis of the thumb screw can neither move up or down as it turns round its axis, the inclination of the threads will rise or fall according to the direction of the thumb screw, and cause the stop to move up and down in the groove on the side of the plane, and thus the stop may be fixed at pleasure. In this plane, the opening for discharging the shaving is upon the right side of the fillister, and in this case the shaving is said by workmen to be thrown on the bench, that is, upon the right side of the plane; but when the orifice of discharge is upon the left, and consequently the shaving thrown upon the left, the plane is said to throw the shaving off the bench; and these expressions are applied to all planes which throw the shavings to one side.

In the moving fillister, as well as in several other planes, the upper part on the sides of the stock is thinner than the lower part; this part is called the hand-hold, and the thick part the body. In the moving fillister, the reduction made for the hand-hold is equally upon both sides of the plane, that is, the rebates are of equal depth. The edges of these rebates, which is the upper surface of the body, are called shoulders; this plane is therefore double shouldered. The same appellation is given to the iron, when a part is taken from one or both sides, so as to make the upper part equally broad, but the sides parallel to the sides of the bottom part. The part of the iron so diminished, is called the tang of the iron, and the broad part at the bottom, which has the cutting edge, is called the web, and the upper narrow surfaces of the web are called the shoulders of the iron, in analogy to those of the plane. The iron of the moving fillister is only single shouldered. Besides the above-mentioned parts, the moving fillister has another, which is a small one-shouldered iron, inserted in a vertical mortise, through the body, between the fore end of the stock and the iron. The web of this little iron is ground with a round basil, from the left side, so as to bring the bottom of the narrow side of the iron to a very convex edge. This little iron is fastened by a wedge, upon the right side of the hand-hold, passing down the mortise in the body. The use of this little iron is principally for cutting the wood transversely when wrought across the fibres, and by this means it not only cuts the vertical side of the rebate quite smooth, but prevents the iron from ragging or tearing the stuff. The whole of this little iron is called a tooth, and the bottom part may be distinguished by the name of the cutter. The cutter must, therefore, stand out a little farther on the right hand side of the plane than the iron, but must never be placed nearer to the fence than the narrow right hand side of the iron. In this plane, the steel side of the iron, and consequently the bedding side of it, is not perpendicular to the vertical sides of the plane, but makes oblique angles therewith, the right hand point of the cutting edge of the iron being nearer to the fore end of the plane than the left hand point of the cutting edge. By this obliquity, the bottom of the

i 2

rebate is cut smoother, particularly in a transverse direction to
the fibres, or where the stuff is cross grained, than could otherwise be done when the steel face of the iron is perpendicular to
the vertical sides of the plane. The principal use is, however,
to contribute, with the form of the cavity, to throw the shaving
into a cylindrical form, and thereby making it issue from one side
of the plane.

----

§ 21. *Of the Sash Fillister in general.* Pl. 12. Fig. 6.

The sash fillister is a rebating plane for reducing the right hand
side of the stuff to a rebate, and is mostly used in rebating the bars
of sashes for the glass, and is therefore called a sash fillister. The
construction of this plane differs in several particulars from the
moving fillister. The breadth of iron is something more than the
whole breadth of the sole, so that the extremities of the cutting
edge are, in a small degree, without the vertical sides of the stock.
In the moving fillister, the fence is upon the bottom of the plane,
and always between the two vertical sides of the stock; but in this
it may be moved to a considerable distance, the limit of which will
be afterwards mentioned. The fence is not moved, as in the moving fillister, by screws fixed in the bottom, but by two bars, which
pass through the two vertical sides of the stock at right angles to
their sides, fitting the two holes exactly through which they pass
in the stock. Each of the bars which thus passes through the
stock, is called a stem, and is rounded on the upper side, for the
convenience of handling. That part of each stem, projecting from
the left hand side of the plane, has a projection downwards, of the
same thickness as the parts which pass through the stock; the bottom sides of these projections are flat surfaces, parallel to the sole
of the plane; the other two sides of the said projections are also
straight surfaces, parallel to the vertical sides of the plane, and are
called the shoulders, so that each stem has three vertical straight
surfaces. The left end of each stem, viz. the end on the left side

of the stock, opposite to the shoulder, may be of any fanciful form. The end of each stem which contains the projection, is called the head of the stem. To each of the heads of the stem, and under each of the lower flat surfaces of the projecting parts, is fixed a piece of wood by iron pins, passing vertically through each head, and through this piece; one of the sides of this piece, next to the stock of the plane, is vertical, and goes about half an inch lower than the sole. The small part of each stem, from the head to the other extremity on the right hand of the stock, is called the tail. The prismatic part is by workmen called the fence. The surface of the fence next to the stock of the vertical plane, and parallel to the vertical faces, is called the guide of the fence. The pins which connect the stem and fence, have their heads on the under side of the fence; the heads are of a conical form; the upper ends of the pins are rivetted upon a brass plate on the round surface of the stem. These pins fix the two stems and the fence stiffly together, but not so much as to prevent either stem from turning round upon the fence, or to make oblique angles with the guide. The upper surface of each stem is rounded, and the two ends ferruled, to prevent splitting when the ends are hit or struck with a mallet, in order to move the guide of the fence either nearer or more remote from the stock, as may be wanted. On the most remote opposite, or vertical sides of the stem, and close to these sides, are cut two small wedge-formed mortises, in which are inserted two small tapering pieces of wood called keys; so that when driven in, or towards the mortise, they will stick fast, and press against the stem, and keep it fast at all points of the tail, and thereby regulate the distance of the fence from the left vertical side of the stock. In order to prevent the keys from being drawn out, or loosing, each has a small elliptic nob at the narrow end, which is also of greater breadth than the mortise upon the left ver tical side of the stock. There are two kinds of sash fillisters, one for throwing the shaving on the bench, and the other for throwing it off: their construction is the same so far as has been described.

§ 22. *The Fillister which throws the Shavings on the Bench,*
PL. 12. FIG. 6.

Has its discharging orifice in course upon the right hana vertical side of the stock, and the left extremity of the cutting edge of
the iron is nearer to the fore end of the plane, than the right hand
extremity of the said edge.   On the left side of the stock, and
from the sole, is a rebate, the depth of which is equal to the depth
of the rebate made on the stuff.   The upper side of the fence
ranges exactly with the side of the rebate which is parallel to the
sole of the plane ; and by this means, the guide of the fence may
be brought quite close to the vertical side of the rebate, or as far
upon the side of the rebate, parallel to the sole of the plane, as may
be found necessary.   The depth of the rebate to be made in the
stuff, is regulated by a stop, which coincides vertically with the
vertical side of the rebate ; the guide of the stop is parallel to the
sole of the plane, and the stop is moved up and down by a thumb
screw, in the same manner as that of the moving fillister, but not
in a groove on the side of the plane, but in a mortise : the side of
the rebate parallel to the sole of the plane, is mortised upwards,
that the guide may be screwed up so as to be flush with that side
of the rebate.   The iron of this plane is single shouldered, and
the projection of the web at the bottom, beyond the tang, is on
the right hand side of the plane, and consequently the narrow
side of the tang and web parts of the iron are in the same straight
line.

---

§ 23. *Of the Sash Fillister for throwing the Shavings*
off the Bench.

The sash fillister wnich throws the shavings off the bench, differs only from the last, in having no rebate on the left hand side
of the plane ; the stop slides in a vertical groove on the left hand
verticle side of the stock, in the same manner as the stop of the
moving fillister, and not in a vertical mortise cut in the vertical

side of the body of the plane : it has also a cutter on the left side, in order to cut the vertical side of the rebate clean. One extremity of the cutting edge of the iron, on the right hand side of the plane, is nearer to the fore end than the other; consequently the steel face of the iron makes angles with the vertical sides of the plane the contrary way to the sash fillister, which throws the shavings on the bench.

§ 24. *Rebating Planes without a Fence.*

Rebating planes which have no fence, are of two kinds; in both, the cutting edge of the iron extends the whole breadth of the sole ; and the upper part of the stock is solid on the two vertical sides, but the lower part is open on both sides; the opening increases from the sole regularly upwards, until it comes to a large cavity, which opens abruptly into a curved form on the side next to the fore end of the plane. The web of the iron is equally shouldered on both sides of the tang.

§ 25. *Skew-mouthed Rebating Plane.*

The thickest stocks, or broadest sole planes, of this description, are made with the face of the iron standing at oblique angles with the vertical sides. The right hand extremity of the cutting edge of the iron, stands nearer to the fore end of the plane than the left hand extremity of the said cutting edge, and the large cavity is greater upon the left side of the plane than upon the right. The shaving is therefore thrown off the bench. The use of this plane is not for sinking the rebate, but rather for smoothing the bottom, after the moving fillister, or after the sash fillister, next to the vertical edge of the rebate. In this manner it is used in cleaning the bottom **entirely** of rebates which do not exceed the breadth of its sole ; but **where** the rebate exceeds this breadth, it is only used next to the vertical side of the rebate as before, and the

N

remaining part of the bottom of the rebate is cleaned off with the trying and smoothing planes. When the iron is set at oblique angles to the vertical sides of the plane, the cutting edge of the sole is said to stand askew, that is, at oblique angles with the sides of the plane. This is therefore called a skew rebating plane. The thickness of this rebating plane is about one inch and five eighths.

----

§ 26. *Square-mouthed Rebating Planes.*

The common rebating planes have the steel side of the iron, or the bed, perpendicular to the vertical sides of the stock, and throw the shaving off the bench; the cavity for the discharge of the shaving is much the same as the skew rebating plane; and since the shaving is thrown off the bench, the widest side of the cavity is on the left hand side of the stock, to clean the internal angles of fillets, and the bottoms of grooves, &c.

----

§ 27. *Side Rebating Planes,*

Are those which have their cutting edge on one side of the plane, and discharge the shaving at the other, the lower part of the stock is therefore open upon both sides. The use of this plane is to clean or plane the vertical sides of rebates, grooves, &c: for this purpose, they are made both right and left: a right hand side rebating plane has its cutting edge on the right hand side of the plane, and consequently throws the shaving off the bench, and the contrary of the left hand rebating plane. The side of the plane containing the mouth, is altogether vertical; but the opposite side is only in part so, from the top downwards to something more than half the height, then recessed and beveled with a taper to the sole; the orifice of discharge for the shaving is beveled. The iron stands askew, or at oblique angles with the mouth side, but perpendicular with regard to the sole or top of the plane; the cutting edge stands nearer to the fore end than the opposite edge. The mortise for the wedge of the iron is without a cavity, as in the other rebating planes, and the iron shouldered upon one side. The web is cut sloping to answer the beveling of the stock.

---

§ 28. *The Plough*, Pl. 12. Fig. 8.

Is used in taking away a solid in the form of a rectangular prism, by sinking any where in the upper surface, but not close to the edge, and thereby leaving an excavation or hollow, consisting of three straight surfaces, forming two internal right angles with each other, and the two vertical sides, two external right angles with the upper surface of the stuff. The channel cut is called a groove, but the operation is called grooving or plowing. The plow consists of a stock, a fence, and a stop. There are two kinds of plows, one where the fence and stop is immoveable, and the other which is universal, of which, both fence and stop are moveable, and will admit of eight or ten irons of various breadths, from one eighth of an inch to three fourths. This is what I shall chiefly describe. The fence has two stems with keys and a stop, moved by a thumb screw, as in the moving fillister for throwing the shaving on the bench. The sole of this plane is the bottom narrow side of two vertical iron plates, which are something thinner than the narrowest iron. The wedge and iron are inserted in the same manner as in the rebating planes, the fore end of the hind plate forms the lower part of the bed of the iron, and has a projecting angle in the middle, and the bed side of each angle has an external angle adapted to the same. This prevents the iron from being removed by the resistance of knots or such sudden obstacles: the fore iron plate is cut with a cavity similar to the common rebate planes. The stop is placed between the fence and sole: this plane is in length about seven inches and three eighths, and in depth three inches and five eighths, and the length of each stem eight inches and a half.

§ 29. *Dado Grooving Plane,*

Is a channel plane, generally about three eighths of an inch broad on the sole, with a double cutter and stop, both placed before the edge of the iron which stands askew; it throws the shaving off the bench. The best kind of dado grooving planes have screw stops of brass and iron; the common sort are made of wood, to slide stiffly in a vertical mortise, and are moved by the blow of a hammer or mallet, by striking the head, when the groove is required to be shallow: but when required to be deep, and consequently the stop to be driven back, a wooden punch must be placed upon the bottom of the stop, and the head of the punch struck with the hammer or mallet, until the guide of the stop arrives at the distance from the sole of the plane that the groove is to be in depth: the use of this plane is for tongueing dado at internal angles, for keying circular dado, grooving for library shelves, or working a broad rebate across the fibres.

§ 30. *Moulding Planes*

Are used in forming curved surfaces of many various fanciful prismatic sections, by way of ornament; these surfaces have therefore this property, that all parallel sections are similar figures. Single mouldings or different mouldings in assemblage have various names, according to their figure, combination, or situation; mouldings are formed either by a plane reversed to the intended section, by a fence and stop on the plane, which causes them to have the same transverse section throughout, or otherwise, by several planes adapted as nearly as possible to the different degrees of curvature; this is called working mouldings by hand. All new or fanciful forms are generally wrought by hand, and particularly in an assemblage of mouldings, where it would be too expensive to make planes adapted to the whole section, or to any particular member or members of that section. The length of moulding planes is nine inches and three eighths, and the depth about three inches and three eighths. Mouldings are said to be stuck when formed by planes, and the operation is called sticking. In mouldings, all internal sinkings which have one flat side, and one convex turned side, are called quirks.

———

§ 31. *Bead Plane*

Is a moulding plane of a semi-cylindric contour, and is generally used in sticking a moulding of the same name on the edge, or on the side close to the arrise : when the bead is stuck upon the edge of a piece of stuff, so as to form a semi-cylindric surface to the whole thickness, the edge is said to be beaded or rounded. When a bead is stuck on, and from one edge on the upper surface of a piece of stuff, so that the diameter may be contained in the breadth of that surface, but not to occupy the whole breadth : then the member so formed has a channel or sinking on the farther side called a quirk, and is therefore called bead and quirk. When the edge of a piece of stuff has been stuck with bead and quirk ; then the vertical side turned upwards and stuck from the same edge in the same manner, another quirk will be formed upon this side provided the breadth of this side be equal to that of the bead ; then the curved surface will be three fourths of a cylinder, this is called bead and double quirk or return bead. The fence is of a solid piece with the plane. The guide of the fence is parallel to the sides of the plane, and tangential to the concave cylindric surface, and its lower edge comes about one fourth or three eighths of an inch below the cylindrical part, the other edge of the cylindrical part forms one side of the quirk, and is on a level with the top of the guide of the fence. The other side of the quirk is a vertical straight surface, and reaches as high as the most prominent part of the cylindric surface of the bead. From the upper edge of this flat side of the quirk, and at right angles to the vertical sides of the plane, proceeds the guide of the stop, which prevents the bead from sinking deeper than the semi-diameter of the cylinder, and the guide of the fence prevents the plane from taking more of the breadth than the diameter. When one, two, or more, contiguous semi-cylinders are sunk within the surface of a piece of wood, with the prominent parts of the curved surface of each, in the same surface as that from which they were sunk, this operation is called reeding, being done in imitation of one or a bundle of a reeds, and each little cylinder is called a reed. In this case, the axis of the reed is in the same straight surface : but this is not always the case, they are sometimes disposed round a staff or rod. Bead planes are sometimes so constructed, as to have the fence taken off or on at pleasure, by screws, for the purpose of striking any series of reeds. When the fence is taken off, the two sides form quirks, and are exactly similar and equal to each other.

The least sized bead is about one eighth of an inch, the next $\frac{5}{32}$, the regular progression stands thus : $\frac{1}{8}$ $\frac{5}{32}$ $\frac{3}{16}$ $\frac{1}{4}$ $\frac{5}{16}$ $\frac{3}{8}$ $\frac{1}{2}$ $\frac{5}{8}$ $\frac{3}{4}$ $\frac{7}{8}$, the first two only differ $\frac{1}{32}$, the next three $\frac{1}{16}$, and from $\frac{3}{8}$ to $\frac{7}{8}$ of an inch, they differ by $\frac{1}{8}$ of an inch each, the $\frac{3}{4}$ and $\frac{7}{8}$ inch beads are torus planes as well as bead planes. The torus only differs from the bead in having a fillet upon the outer edge of the stuff : consequently the torus consists of a fillet and semi-cylinder. It may be observed, that whether there be one or two semi-cylinders stuck on the edge of a piece of stuff, that without there is a fillet upon the edge they only take the name of beads. The torus is in general much larger than the bead : but when there are two semi-cylinders with a fillet upon the outer edge, the combination is called a double torus, and if there is no fillet, it is called a double bead, even though the one should be much larger than the other.

———

§ 32. *A Snipesbill*

Is a moulding plane for forming a quirk : snipesbills are of two kinds, one for sinking the quirk, called a sinking snipesbill, and the other for cleaning the vertical flat side of the quirk, called a side snipesbill. Each of these two kinds are right and left.

In the sinking snipesbill the cutting edge is on the sole, and the extremity of the iron comes close to the side of the plane, which forms the vertical side of the quirk ; the sole consists of two parts of a cylindric surface of contrary curvature : one next to the edge which forms the quirk, is concave, and the part more remote, is convex.

The side snipesbill has its iron placed very nearly perpendicular, with regard to the sole of the plane, the top of the iron leaning about five degrees forward : this plane has its cutting edge upon one side or the other, according to the side or to the hand it is made for. The iron stands askew to the vertical sides of the plane.

---

§ 33. *Hollows and Rounds*

Are mouldings for striking convex and concave cylindrical surfaces, or any segment or parts of these surfaces ; they have therefore their soles exactly the reverse of what is intended. Hollows and rounds are not confined to cylindric surfaces, but will also stick those of cylindrical forms, or those which have elliptic sections, perpendicular to the direction of the motion by which they are wrought. Mouldings depressed within the surface of a piece of wood, or those which form quirks, must first be sunk by the snipesbill, and formed into the intended shape by hollows and rounds. The hollow is only used in finishing a convex moulding; the rough is generally taken off with the jack plane, when there is room to apply it, if not, with the firmer chisel. In making a hollow, a rough excavation is first made with a gouge, and then finished with the round, and sometimes with two rounds, of which the sole of the one that comes first is a little quicker, and the iron set more rank.

§ 34. *Stock and Bits*, Pl. 13. Fig. 8.

The stock is a wooden lever, to be turned round an axis swiftly by hand, in order to give the same rotative motion round the axis, to a piece of steel fixed in the said axis, the steel being sharpened at the extremity, so as to cut a cylindric hole, in the same direction as the axis of the stock.

The axis is continued on both sides of the handle or winch part; one part of the axis is made with a broad head, to be placed against the breast while boring, even when pressing pretty hard upon the stock, and is so constructed with a joint, as to be stationary, while all the other parts are in motion; the lower part of the stock is brass, and is fixed to it by means of a screw passing through two ears of the brass part, and through the solid of the wood. The brass part is called the pad, which is so contrived, as to admit of different pieces of steel called bits, for boring and widening holes of various diameters in wood, and countersinking, both in wood and iron; that is, forming a cavity or hollow cone on the outer side of a cylindric hole to receive the head of a screw, or the like. The upper part of each bit inserted in the stock, is the frustum of a square pyramid, which goes into a hollow mortise of the same form, and is secured by means of a spring fixed in the pad, and which falls into a notch at the upper end of the bit.

The construction of bits depends upon their use. Small bits are used for boring of wood, and have an interior cavity for containing the core, separated from the wood by the under edge. The lower part of the cavity is the surface of a cylinder, and the upper part where the cavity ends is a part of a long hollow oblong spheroid, terminated upon the sides of the bit: the exterior side is also cylindrical, as high as that of the interior, and thence diminishes for a considerable way above the hollow, that it may turn in the hole with the greater ease. The section of the bit is the figure of a crescent. The cutting edge has its basil on the inside, and stands prominent in the middle; this bit is also called a pin or gouge bit from its being mostly used in framing : it bores soft wood, as deal, with greater rapidity than any other tool.

---

§ 35. *The Centre Bit*

Is constructed with a projecting conical point nearly in the middle, called the centre of the bit; on the narrow vertical surface, the one most remote from the centre is a tooth with a cutting edge. The under edge of the bit on the other side of the centre, has a projecting edge inclined forward. The horizontal section of this bit upwards is a rectangle. The axis of the small cone in the centre is in the same straight line as that of the stock; the cutting edge of the tooth is more prominent than the projecting edge on the other side of the centre, and the vertex of the conic centre still more prominent than the cutting edge of the tooth.

The use of the centre bit is to form a cylindric excavation, having the upper point of the axis of the intended hole, given on the surface of the wood : the centre of the bit is first fixed in this point, then placing the axis of the stock and bit in the axis of the intended hole to be bored, with the head of the stock against your breast, lay hold of the handle and turn the stock swiftly round, then the hollow cone made by the centre will cause the point of the tooth to move in the circumference of a circle, and cut the cylindric surface progressively as it is turned round, and the projecting edge upon the other side of the centre, will cut out the core in a spiral formed shaving : centre bits are of various sizes, in order to accommodate bores of different diameters.

---

§ 36. *Countersinks*

Are bits for widening the upper part of a hole in wood or iron, for the head of a screw or pin, and have a conical head. Those for wood have one cutter in the conic surface, and have the cutting edge more remote from the axis of the cone than any other part of

No. 8.   o

the surface. Countersinks for brass have eleven or twelve cutters round the conic surface, so that the horizontal section represents a circular saw. These are called rose countersinks. The conic angle at the vertex is about ninety degrees. Countersinks for iron have two cutting edges, forming an obtuse angle.

---

§ 37. *Rimers*

Are bits for widening holes: for this purpose they are of a pyramidical structure, having their vertical angle about three degrees and a half. The hole must first be pierced by means of a drill or punch; when the rimer is put into the stock, and the point into the hole, and being turned swiftly round, the edges will cut or scrape off the interior surface of the hole as it sinks downwards, by pressing upon the head of the stock. Brass rimers have their horizontal sections of a semicircular figure, and those for iron polygonal: of these some have their sections square, some hexagonal, and some octagonal.

---

§ 38. *The Taper Shell Bit*

Is conical both within and without, and the horizontal section a crescent, the cutting edge is the meeting of the exterior and interior conic surface. The use of this bit is for widening holes in wood.

Besides the above bits, some stocks are provided with a screw driver for sinking small screws into wood with greater rapidity than could be done by hand.

---

§ 39. *The Brad Awl*, Pl. 13. Fig. 3.

Is the smallest boring tool, its handle is the frustum of a cone tapering downwards. The steel part is also conical, but tapering upwards, and the cutting edge is the meeting of two basils, ground equally from each side. A hole is made by placing the edge transverse to the fibres of the wood, and pushing the brad awl into the wood, turning it to and fro by a reciprocal motion. The core is not brought out as by the other boring instruments; but the wood is displaced and condensed around the hole. Brad awls are used for making a way for brads, and are of several sizes; they are not so apt to split the wood as the gimlet.

---

§ 40. *Chisels in general.* PL. 13. FIGS. 3, 4, 5.

A chisel is an edge tool for cutting wood, either by leaning on it, or by striking it with a mallet. The lower part of the chisel is the frustum of a cuneus or wedge, the cutting edge is always on, and generally at right angles to the side. The basil is ground entirely from one side. The two sides taper in a small degree upwards, but the two narrow surfaces taper downwards in a greater degree. The upper part of the iron has a shoulder, which is a plain surface at right angles to the middle line of the chisel. From this plain surface rises a prong in the form of a square pyramid, the middle line of which is the same as the middle line of the cuneus or wedge: the prong is inserted and fixed in a socket of a piece of wood of the same form. This piece of wood is called the handle, and is generally the frustum of an octagonal pyramid, the middle line of which is the same as that of the chisel; the tapering sides of the handle diminish downwards, and terminate upwards in an octagonal dome. The use of the shoulder is for preventing the prong from splitting the handle while being struck with the mallet. The chisel is made stronger from the cutting edge to the shoulder, as it is sometimes used as a lever, the prop being at or near the middle, and the power at the handle, and the resistance at the cutting edge; some chisels are made with iron on one side, and steel on the other, and others consist entirely of steel.

There are several kinds of chisels, as the paring chisel, the mortise chisel, the socket chisel, and the ripping chisel.

§ 41. *The Firmer Chisel*, Pl. 13. Fig. 4.

Is used both by carpenters and joiners in cutting away the
superfluous wood by thin chips. The best are made of cast steel.

When there is a great deal of superfluous wood to be cut away,
sometimes a strong chisel consisting of an iron back and steel face
is first used, by driving it into the wood with a mallet, and then a
slighter one, consisting entirely of steel sharpened to a very fine
edge, is used in the finish. The first used is called a firmer, and
the last, a paring chisel, in working which, only the shoulder or
hand is employed in forcing it into the wood.

---

§ 42. *The Mortise Chisel*, Pl. 13. Fig. 5.

Is made exceedingly strong, for cutting out a rectangular prismatic cavity across the fibres, quite through or very deep in a
piece of wood, for the purpose of inserting a rectangular pin of the
same form on the end of another piece of wood, and thereby fastening the two pieces of wood together. The cavity is called a
mortise, and the pin inserted a tenon : and the chisel used for
cutting out the cavity is therefore called a mortise chisel. As
the thickness of this chisel from the face to the back is great, in
order to withstand the percussive force of the mallet ; and as the
angle which the basil makes with the face is about twenty-five
degrees, the slant dimension of the basil is very great. This chisel
is only used by percussive force, given by the mallet.

---

§ 43. *The Gouge*

Is used in cutting an excavation of a concave form, and is similar
to the chisel, except that the bottom part is cylindrical both within
and without ; the basil is made on the inside ; the best are those
which are made of cast steel.

§ 44. *The Drawing Knife*

Is an oblique ended chisel, or old knife, for drawing in the ends of tenons, by making a deep incision with the sharp edge, by the edge of the tongue of a square : for this purpose a small part is cut out in the form of a triangular prism, and consequently the hollow will contain one interior angle and two sides, one side next the body of the wood being perpendicular, and the other inclined. The use of this excavation is to enter the saw, and keep it close to the shoulder, and to make the end of the rail quite smooth, for the saw will not only be liable to get out of its course into a new direction, but may tear and scratch the wood at the shoulder.

---

§ 45. *Of Saws in general.* PL. 13. FIG. 6, 7, 8, 9, 13.

A saw is a thin plate of steel indented on the edge for cutting, by a reciprocal change in the direction of motion, pushing it from, and drawing it towards you. The cut which it makes, or the part taken away in a board, is a thin slice, contained between parallel planes, or a deep narrow groove of equal thickness. Saws are of several kinds, as the ripping saw, the half ripper, the hand saw, the panel saw, the tenon saw, the sash saw, the dove-tail saw, the compass saw, and the key-hole or turning saw. The teeth of these saws are all formed so as to contain an angle of sixty degrees, both external and internal angles, and incline more or less forward as the saw is made to cut transverse to, or in the direction of the fibres : they are also of different lengths and breadths, according to their use. The teeth of a saw are bent alternately to each side, that the plate may clear the wood.

---

§ 46. *The Ripping Saw*

Is used in dividing or slitting wood in the direction of the fibres; the teeth are very large, there being eight in three inches, and the front of the teeth stand perpendicular to the line which ranges with the **points**: the length of the plate is about twenty eight inches.

---

§ 47. *The Half Ripper*

Is also used in dividing wood in the direction of the fibres: the .cngth of the plate of this is the same as the former, but there are only three teeth in the inch.

---

§ 48. *The Hand Saw*, Pl. 13. Fig. 6.

Is both used for cutting the wood in a direction of the fibres and cross cutting: for this purpose the teeth are more reclined than the two former saws: there are fifteen teeth contained in four inches. The length of the plate is twenty six inches.

---

§ 49. *The Panel Saw*

Is used for cutting very thin wood, either in a direction of, or transverse to the fibres. The length of the plate is the same as that of the hand saw, but there are only about six teeth in the inch. The plates of the hand saw and panel saw are thinner than the ripping saw.

---

§ 50. *The Tenon Saw*, Pl. 13. Fig. 7.

Is generally used for cutting wood transverse to the fibres, as the shoulders of tenons. The plate of a tenon saw is from fourteen to nineteen inches in length, and the number of teeth in an inch from eight to ten. As this saw is not intended to cut through the wood its whole breadth, and as the plate would be too thin to make a straight kerf, or to keep it from buckling, there is a thick piece of iron fixed upon the other edge for this purpose, called the back.

The opening through the handle for the fingers of this and the foregoing saws is inclosed all round; and on this account is called a double handle.

—

§ 51. *The Sash Saw*, PL. 13. FIG. 8.

Is used by sash makers in forming the tenons of sashes: the plate is eleven inches in length. The inch contains about thirteen teeth; this saw is sometimes backed with iron, but more frequently with brass.

—

§ 52. *The Dove-tail Saw*

Is used in dove-tailing drawers. The length of the plate is about nine inches, and the inch contains about fifteen teeth. This plate is also backed with brass. The handles of the two last saws are only single.

—

§ 53. *The Compass Saw*, PL. 13. FIG. 9.

Is for cutting the surfaces of wood into curved surfaces: for this purpose it is narrow, without a back, thicker on the cutting edge, as the teeth have no set. The plate is about an inch broad, next to the handle, and diminishes to about one quarter of an inch at the other extremity; here are about five teeth in the inch. The handle is single.

—

§ 54. *The Key-hole, or Turning Saw*

Is similar to the compass saw in the plate, but the handle is long, and perforated from end to end, so that the plate may be inserted any distance within the handle. The lower part of the handle is provided with a pad, through which is inserted a screw, for the purpose of fastening the plate in the handle. this saw is

used for turning out quick curves, as key-holes, and is therefore
frequently called a key-hole saw.

___

§ 55. *The Hatchet*

Is a small axe, used chiefly in cutting away the superfluous
wood from the edge of a piece of stuff, when the part to be cut
away is too small to be sawed.

___

§ 56. *The Square,* Pl. 13. Fig. 11.

Consists of two rectangular prismatic pieces of wood, or one of
wood, and the other which is the thinest, of steel, fixed together,
each at one of their extremities, so as to form a right angle both
internally and externally; the interior right angle is therefore
called the inner square, and the exterior one the outer square.
The side of the square which contains the mortise, or through
which the end of the other piece passes, is made very thick, not
only that it may be strong enough for containing the tenon of the
other piece, but that it should keep steady and flat when used;
and the piece which contains the tenon is made thin, in order to
observe more clearly whether the edge of the square and the
wood coincide. The thick side of the square is called the stock
or handle, and the narrow surface of the handle is always applied
to the vertical surface of the wood. The thin side of the square
is called the blade, and the inner edge of the blade is always
applied to the horizontal surface of the wood. Squares are of
different dimensions according to their use: some are employed
in trying-up wood, and some for setting out work; the former is
called a trying square, and the latter a setting-out square; the
blade ought to be of steel, and always ought to project beyond the
end of the stock, particularly if made of wood. The stock is
always made thick, that it may be used as a kind of fence in keeping the blade at right angles to the arris.

§ 57. *To prove a Square.*

Take a straight edged board which has been faced up, and apply the inner edge of the stock of the square to the straight edge of the board, laying the side of the tongue upon the face of the board; with a sharp point draw a line upon the surface of the board by the edge of the square: turn the square so that the other side of the blade may lie upon the face of the board; bring the stock close to the straight edge of the board, then if the edge of the square does not lie over the line, or any part of the line, the square must be shifted until it does, then if the edge of the tongue of the square and the line coincide, the square is already true: but if there is an open space between the farther side of the board and the straight edge, that is, if the farther end of the edge of the tongue of the square meets the farther end of the line from the straight edge, draw another line by the edge of the tongue of the square, and these two lines will form an acute angle with each other, the vertex of which will be at the farther side of the board, and the opening towards the straight edge: take the middle of the distance between the two lines at the arris, and draw a line from the middle point to the point of concourse of the lines: then the blade of the square must be shot or made straight, so as to coincide with this last line. The same, or a similar operation, must be repeated, if the contrary way.

----

§ 58. *The Bevel,* Pl. 13. Fig. 12.

Consists of a blade and handle the same as the square, except that the tongue is made moveable on a joint that it may be set to any angle. When many pieces of stuff are to be tried up to a particular angle, an immoveable bevel ought to be made for the purpose, for unless very great care be taken in laying down the moveable bevel, it will be liable to shift.

P

§ 59. *The Gauge*, Pl. 13. Fig. 13.

Is an instrument for drawing a line parallel to the arris of a piece of stuff, on one or both of the adjoining surfaces. It consists of a thick rectangular prismatic part, with a mortise of the same figure, cut perpendicularly through it, between two of its opposite sides, and this prism is called the head. In the mortise is inserted another prism exactly made to fill its cavity, this prism is called the stem; at one end of the stem is a steel tooth projecting perpendicularly from the surface, so that by striking one end or other with the mallet, the tooth is moved farther or nearer to the adjacent surface of the head, as the distance may be wanted be tween the arris of the stuff and the line to be marked out by the tooth.

------

§60. *The Mortise Gauge*

Is constructed similar to the common gauge, but has two teeth instead of one. One tooth is stationary at the end of the stem, and the other is moveable in a mortise between the fixed tooth and the head, so that the distances of the teeth from each other, and of each tooth from the head, may be set in any ratio or proportion to each other, that the thickness of a tenon or wood may require. The use of this gauge is, as its name implies, for gauging mortises and tenons.

------

§ 61. *The Side Hook*, Pl. 12. Fig. 11.

Is a rectangular prismatic piece of wood with two projecting knobs upon the alternate sides of it. Every joiner ought to be provided with at least two side hooks of equal size. Their use is to hold a board fast, the fibres of the board running in the direction of the length of the bench, while the workman is cutting across the fibres with a saw or grooving plane, or in traversing the wood, which is planing in a direction perpendicular to the fibres, or with very little obliquity.

§ 62. *The Mitre Box*

Is used for cutting a piece of tried-up stuff at an angle of forty-five degrees with two of its surfaces, or at least to one of the arrises, and perpendicular to the other two sides, or at least to one of them obliquely to the fibres. The mitre box consists of three boards, two called, the sides being fixed at right angles to the third, the bottom : the bottom and top of the sides are all parallel : the sides are of equal height, and cut with a saw into two directions of straight surfaces at right angles to each other and to the bottom, forming an angle of forty-five degrees with the sides.

---

§ 63. *The Shooting Block*

Is two boards fixed together, the sides of which are lapped upon each other, so as to form a rebate for the purpose of making a short joint, either oblique to the fibres, or in their direction. By this instrument the joints of panels for framing are made, also the joints for the mitres of architraves, or the like.

---

§ 64. *The Straight Edge*

Is a piece of stuff or board made perfectly straight on the edge, in order to make other edges straight, or to plane the face of a board straight.

Straight edges are of different dimensions, as the magnitude of the work may require.

---

§ 65. *Winding Sticks*

Are two pieces of wood of equal breadth for the purpose of ascertaining whether a surface be straight or not ; if not, the surface must be brought to a straight by trial.

§ 66. *The Mitre Square*

Is so called, because it bisects the right angle, or mitres the square, and is therefore an immoveable bevel, made to strike an angle of forty-five degrees with one side or edge of a piece of stuff, upon the adjoining side or edge of the said piece of stuff: it consists of a broad thin board let in, or tongued into a piece on the edge, called the fence or handle; the fence projects equally upon each side of the thin piece or blade, of which one of the edges is made to contain an angle of forty-five degrees with the nearest edge of the handle, or of that in which the blade is inserted. The inside of the handle is called the guide; the handle may be about an inch thick, two inches broad, the blade about a quarter of an inch, or about one eighth and a sixteenth. The blade may be about seven or eight inches broad; but mitre squares must be of various sizes, according to the work, and consequently of different thicknesses.

To use the mitre square, lay the guide of the handle upon the arris, slide it along the stuff until the oblique edge comes to the place required, then draw a line by this edge; the angle of the mitre may be struck either way, according to the direction required, by turning the mitre square.

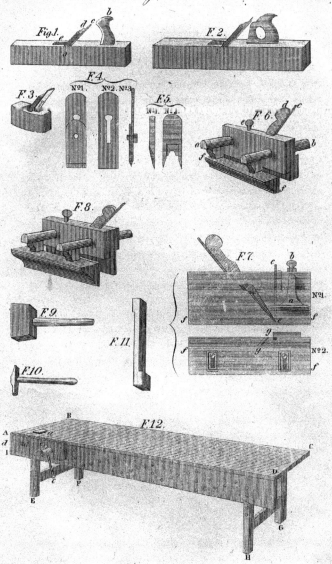

Joinery Plate XII.

§ 67. EXPLANATION OF THE PLATES IN JOINERY.

PLATE XII.

TOOLS.

Fig. 1 the jack plane, *a* the stock, *b* the tote or handle, being a single tote, *c* the iron, *d* the wedge for tightening the iron, *e* the orifice or place of discharge for the shavings.

Fig. 2 the trying plane, the parts are the same as the jack plane, except that the hollow of the tote is surrounded with wood, and is therefore called a double tote.

Fig. 3 is the smoothing plane without a tote, the hand-hold being at the hind end of the plane.

Fig. 4 the iron, No. 1. the cover for breaking the shaving, screwed upon the top of the iron, in order to prevent the tearing of the wood, in a front view : No. 2. front of the iron without the cover, showing the slit or the screw which fastens the cover to the iron: No. 3. profile of iron and cover screwed together.

Fig. 5 the wedge for tightening the iron : No. 1. longitudinal section of the wedge : No. 2. front, showing the hollow below for the head of the screw.

Fig. 6 sash fillister, for throwing on the bench, *a* head of one stem, *b* tail of the other, *c* iron, *d* wedge, *e* thumb screw for moving the stop up and down, *f f* fence for regulating the distance of the rebate from the arris.

Fig. 7 the moving fillister for throwing the shaving on the bench : No. 1. right hand side of the plane, *a* brass stop, *b* thumb screw of do. *c d e* tooth, the upper part *c d* on the outside of the neck, and the part *d e* passing through the solid of the body with a small part open above, *e*, for the tang of the iron tooth, *f f* the guide of the fence : No. 2. bottom of the plane turned up, *a* the guide of the stop, *f f* the fence, showing the screws for regulating the guide, *g g* the mouth and cutting edge of the iron.

Fig. 8 the plow, the same with regard to the stem fence and stop, and also in other respects as the sash fillister, except the sole, which is a narrow iron.

Fig. 9 the mallet.

Fig. 10 the hammer.

Fig. 11 the side hook for cutting the shoulders of tenons.

Fig. 12 the work bench, *a* the bench hook, *b b* the screw check, *c c* handle screw, *d* end of guide.

PLATE XIII.

TOOLS.

Fig. 1 stock, into which is fixed a centre bit.

Fig. 2 No. 1. the gimlet: No. 2. the lower part at full size.

Fig. 3 No. 1. the brad awl: No. 2. the lower end turned edge-ways: No. 3. the lower end turned side-ways.

Fig. 4 No. 1. the paring chisel: No. 2. the lower end turned edge-ways with the basil.

Fig. 5 the mortise chisel: No. 1. side of the chisel: No. 2. front: No. 3. lower end with the basil.

Fig. 6 hand saw.

Fig. 7 tenon saw, with back generally of iron.

Fig. 8 sash saw, backed generally with brass.

Fig. 9 compass saw for cutting curved pieces of wood.

Fig. 10 key hole saw, *a* the pad in which are inserted a spring and two screws, for fixing the saw to any length.

N. B. The hand saw and tenon saw have what are called double handles, and the tenon and compass saws single handles. The position and form of the handle depends on the position of the working direction of the saw.

Fig. 11 the square, *a b c* the outer square, *d e f* the inner square, *a d e* the stock or handle, *b c f e* the blade.

Fig. 12 the moveable bevel, *a b* the stock, *b c* the blade.

Fig. 13 the gauge, *a a* the stem, *b b* the head which moves, *c* the tooth which marks.

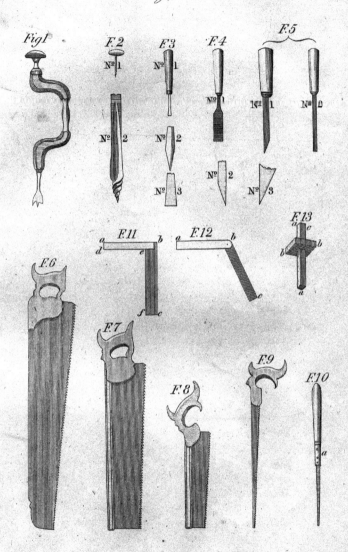

Fig 1 F.2 F.3 F.4 F.5

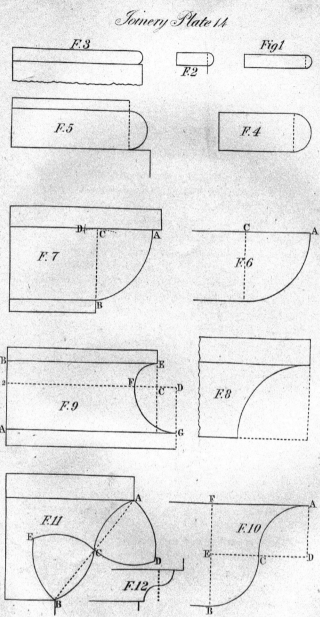

Joinery Plate 14

PLATE XIV.

MOULDINGS.

§ 68. *To draw the several kinds of Mouldings made by Joiners.*

An astragal is a moulding of a semi-circular profile; its con-struction is so simple that it would be unnecessary to say any thing concerning it. Fig. 1.

There are two kinds of beads, one is called a cocked bead, when it projects beyond the surface to which it is attached, see Fig. 2; and the other is called a sunk bead, when the sinking is depressed beneath the surface of the material to which it is at-tached, that is, when the most prominent part of the bead is in the same surface with that of the material, Fig. 3.

A torus in architecture is a moulding of the same profile as a bead; the only difference is, when the two are combined in the same piece of work, the torus is of greater magnitude, as Fig. 4; in Joinery the torus is always accompanied with a fillet. Fig. 5. single torus moulding.

The Roman ovolo or quarter round, as called by joiners, is the quadrant of a circle, Fig. 6. When the projection and height are unequal, as in Fig. 7, take the height B C, and from the point B describe an arc at C, and with the same radius from A, describe another arc cutting the former at D, with the distance A D or D B describe the profile A B. This is generally accompanied with fillets above and below, as in Fig. 7.

The cavetto is a concave moulding, the regular profile of which is the quadrant of a circle, Fig. 8; its description is the same as the ovolo.

A scotia is a concave moulding receding at the top, and pro-jecting at the bottom, which in this respect is contrary both to the ovolo and cavetto; it is also to be observed, that its profile consists of two quadrants of circles of different radii, or it may be consi-dered as a semi-ellipse taken upon two conjugate diameters, Fig. 9.

To describe the scotia, divide the height A B into three equal parts, at the point 2 draw the line 2 C D, being one-third from the top, draw E C perpendicular to C D, with the centre C and and distance C E describe the quadrant E F; take the height A 2 and make F D equal to it: draw D G perpendicular to F D, from D with the distance D F describe the arc F G, and E F G will be the profile of the scotia. This moulding is peculiarly applied to the bases of columns, and makes a distinguishing line of shadow between the torii.

The ogee is a moulding of contrary curvature, and is of two kinds: when the profile of the projecting part is concave, and consequently the receding part convex, the ogee is called a cima-recta, Figs. 10 and 11; and when the contrary, it is then called a cima-reversa, Fig. 12.

To describe the cima-recta when the projection of the moulding is equal to its height, and when required to be of a thick curvature, Fig. 10. Join the projections of the fillets A and B by the straight line A B; bisect A B at C, draw E C D parallel to the fillet F A, draw A D and B E perpendicular to F B; from the point E describe the quadrant B C, and from the point D describe the quadrant A C, then B C A is the profile.

To describe the cima-recta when the height and projection are unequal, and when it is required to be of a flat curvature, Fig. 11. Join A B and bisect it in C, with the distance B C or C A from the point A describe the arc C D, from C with the same radius describe the arc A D cutting the former in D, the foot of the compass still remaining in C describe the arc B E, from B with the same radius describe the arc C E, from the point D describe the arc A C, from the point E describe the arc C B, then will A C B be the profile required.

The cima-reversa, Fig. 12, is described in the same manner.

Quirk mouldings sometimes occasion confusion as to their figure particularly when removed from the eye, so as frequently to make one moulding appear as two.

Fig.1.

F.2.

F.3.

F.4.

F.5.

F.6.

F.7.

F.8.

F.9.

F.10.

F.11.

F.12.

F.13.

PLATE XV.

§ 69. MOULDINGS.

The names of mouldings according to their situation and combination, in various pieces of joiners' work.

Fig. 1 edge said to be rounded.

Fig. 2 quirked bead or bead, and quirk.

Fig. 3 bead and double quirk, or return bead.

Fig. 4 double bead, or double bead and quirk.

Fig. 5 single torus.

Fig. 6 double torus. Here it is to be observed, that the distinction between torus mouldings and beads in joinery is, the outer edge of the former always terminates with a fillet, whether the torus be double or single, whether in beads there is no fillet on the outer edge.

Figs. 7, 8, 9 single, double, and triple reeded mouldings; semicylindric mouldings are denominated reeds, either when they are terminated by a straight surface equally protuberant on both sides, as in these figures, or disposed longitudinally round the circum ference of a shaft ; but if only terminated on one side with a flush surface, they are then either beads or torus mouldings.

Fig. 10 reeds disposed round the convex surface of a cylinder.

Figs. 11, 12, 13 fluted work. When the flutes are semi-circular, as in Fig. 11, it is necessary that there should be some distance between them, as it would be impossible to bring their junction to an arris ; but in flutes, the sections of which are flat segments, the flutes generally meet each other without any intermediate straight surface between them. The reason of this is, that the light and shade of the adjoining hollows are more contrasted, the angle of their meeting being more acute, than if a flat space were formed between them. See Figs. 12 and 13, fluting round the convex surface of a cylinder.

PLATE XVI.

§ 70. *Mouldings of Doors, &c.*

The different denominations of framed doors, according to their mouldings and panels, and framed work in general. The figures in the plates to which these descriptions refer, are sections of doors, through one of the stiles taking in a small part of the panel, or they may be considered as a vertical section through the top rail, showing part of the panel.

Fig. 1 the framing is without mouldings, and the panel a straight surface on both sides : this is denominated doors square and flat panel on both sides.

Fig. 2 the framing has a quirked ovolo, and a fillet on one side, but without mouldings on the other, and the panel flat on both sides : this is denominated doors quirked ovolo, fillet and flat, with square back.

Fig. 3 differs only from the last in having a bead instead of a fillet, and is therefore denominated quirked ovolo bead and flat panel, with square back.

Fig. 4 has an additional fillet on the framing, to what there is in Fig. 3, and is therefore denominated quirked ovolo bead, fillet and flat panel, with square back.

Note. When the back is said to be square, as in Figs. 2, 3, 4, the meaning is, that there are no mouldings on the framing, and the panel is a straight surface on one side of the door.

Fig. 5 the framing struck with quirk ogee and quirked bead on one side, and square on the other; the surface of the panel straight on both sides : this is called quirked ogee, quirked bead, and flat panel, with square back.

Fig. 6 differs from the last only in having the bead raised above the lower part of the ogee and a fillet.   This is therefore denominated quirked ogee, cocked bead, and flat panel, with square back.

Joinery   Plate 16.

Fig. 1.

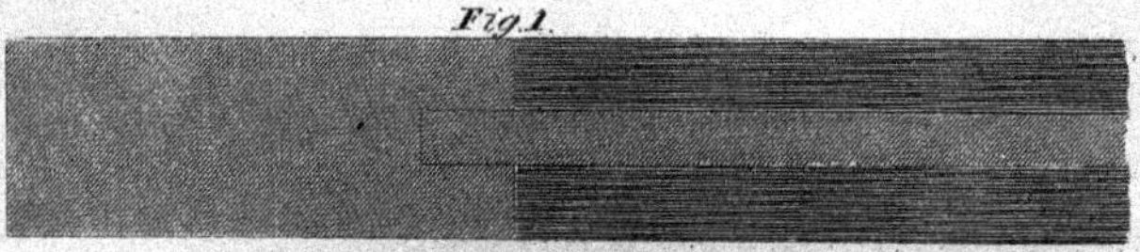

F. 2.

F. 3.

F. 4.

F. 5.

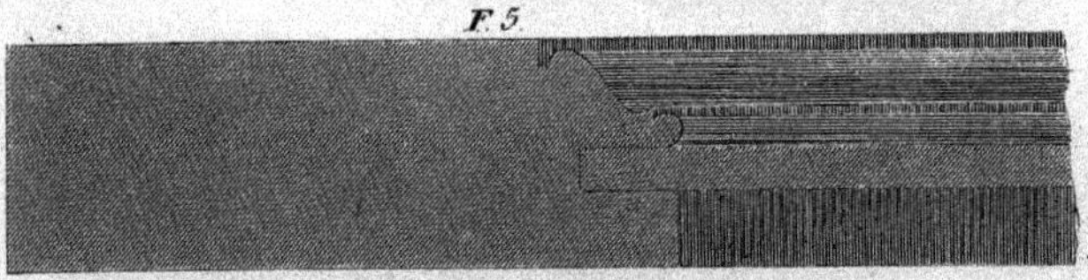

F. 6.

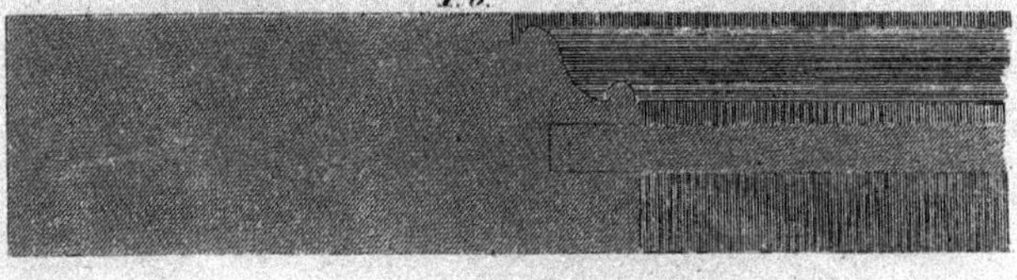

Fig. 1.

F. 2.

F. 3.

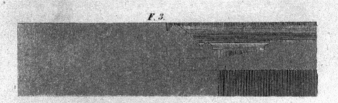

F. 4.

F. 5.

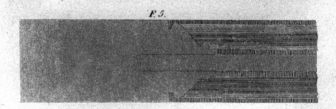

PLATE XVII.

Mouldings for Doors, &c.

Fig. 1 is denominated cove, cocked bead, and flat panel, with square back.

Fig. 2 is denominated quirked ovolo, bead, fillet, and raised panel on front, with square back. The rising of the panel gives strength to the door, and on this account they are often employed in street doors, though the fashion at present is discontinued in the inside of building.

Fig. 3 the framing is the same as the last, but the panel is raised in front, and has an ovolo on the rising. This is therefore denominated quirked ovolo, bead, and raised panel, with ovolo on the rising on front of door, with square back.

Fig. 4 is denominated quirked ogee, raised panel, ovolo, and fillet on the rising and astragal on the flat of panel in front and square back.

Note. The raised side of the panel is always turned towards the street.

Fig. 5 is denominated quirked ovolo, bead, fillet, and flat panel, on both sides; doors of this description are used between rooms, or between passages and rooms, where the door is equally exposed on both sides. When the panels are flat on both sides, or simply chamfered on one side and flat on the other, and the framing of the door moulded on the side which has the flat panels; such doors are employed in rooms where one side only is exposed, and the other never but when opened, being turned towards a cupboard or dark closet.

PLATE XVIII.

Mouldings for Doors, &c.

Fig. 1 is denominated bead, but, and square, or more fully bead and but, front and square back. In bead and but work, the bead is always struck on the outer arris of the top or flat of the panel in the direction of the grain.

Fig 2 is denominated bead and flush front and quirked ogee, raised panel, with ovolo on the rising, grooved on flat of panel, on back. Bead and flush, and bead and but work are always used where strength is required. The mouldings on the inside are made to correspond with the other passage or hall doors.

Fig. 3 is a collection or series of mouldings the same on both sides, and project in part without the framing on each side; the mouldings are laid in after the door is framed square and put together. If braded through the sides of the quirks, the heads will be entirely concealed; but observe, that the position of the brads must not be directed towards the panels, but into the solid of the framing. The mouldings of doors which thus project are termed belection mouldings; belection moulded work is chiefly employed in superior buildings.

Fig. 4 another form of a belection moulding.

The following is a geometrical description of reeded mouldings, sash bars, and the manner of springing mouldings.

Fig. 5 to inscribe a circle in a given sector A B C of a circle, bisect the angle B A C by G A; produce the sides A B, A C, to D and E, and A G to meet the arc in F, draw D E perpendicular to A F, bisect the angle D E A of the triangle A D E by E G, and G is the centre of the inscribed circle, and G F the radius.

Fig. 6 a reeded staff, the reeds described as in Fig. 5

Fig. 1.

F. 2.

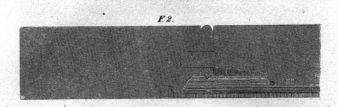

F. 3.

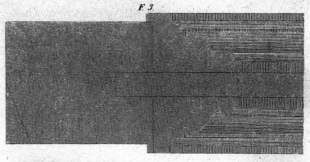

F. 4.

F. 5.

F. 6.

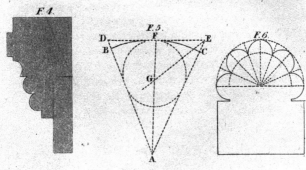

Joinery Plate XIX.

Fig.1   F.2   F.3   F.4

F.5   F.6   F.7   F.8

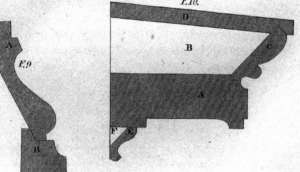

F.10.

D

B

C

A

A

F.9

F E

B

PLATE XIX.

Mouldings for Sashes and Cornices.

Fig. 1 simple astragal or half round bar for sashes.

Fig. 2 quirked astragal bar.

Fig. 3 quirked Gothic bar.

Fig. 4 another form of a Gothic bar.

Fig. 5 double ogee bar: this and the preceding forms are easily kept clean.

Fig. 6 quirked astragal and hollow: bars of this structure have been long in use.

Fig. 7 double reeded bar.

Fig. 8 triple reeded bar.

Fig. 9 base moulding of a room with part of the skirting. When the base mouldings are very large, they ought to be sprung as in this diagram. A the base moulding, B part of the plinth. In order to know what thickness it would require a board to be of, to get out a moulding upon the spring; the best method is to draw the moulding out to the full size, then draw a line parallel to the general line of the moulding, so as to make it equally strong throughout its breadth, and also of sufficient strength for its intended purpose.

Fig. 10 a cornice. The part A forming the corona, is got out of a plank. B is a bracket, C the moulding on the front spring, D a cover board forming the upper fillet, E a moulding sprung below the corona, F a bracket.

---

§ 71. *Definitions.*

A piece of stuff is said to be wrought when it is planed on one or more sides, so as to make a complete finish as far as required by a plane; hence if it is only planed with the jack plane, and no farther operation of any other plane required, in this case it is

L 2

said to be wrought; and if the stuff requires to be made straighter with the trying plane, the stuff is still said to be wrought.

The operation of planing the first side of a board or piece of stuff straight, is called facing, the side so done is called the face, and the board itself is said to be faced-up.

The operation of planing the edge of a board straight, is called shooting, and the edge is said to be shot.

When two adjoining surfaces of a piece of stuff are planed so as to form a right angle, the piece of stuff is said to be squared.

When two adjoining surfaces of a piece of stuff are planed so as to form an acute or obtuse angle by the inclination of these surfaces, this piece of stuff is said to be bevelled; and if one surface is narrower than the other, the narrower surface becomes the edge, the edge is then said to be bevelled; but this is only meant in reference to the face, as the expression could have no meaning, except in the relation of the adjoining surfaces. The same is also applied to a piece of wood that has been squared, the edge is said to be squared, instead of the adjoining surfaces said to be squared.

When a line has been drawn on the face or edge of a piece of stuff parallel to the arris or line of concourse of the two surfaces that are planed, that surface is said to be gauged, and is generally done by means of the implement or tool called a gauge.

When the stuff is planed on one, two, three, or all the four sides, as may be required, then the stuff is said to be tried up; the term try-up is sometimes applied to facing, but in what follows, the term facing, is only applied to the side first wrought.

––––––

§ 72. *To make a Straight Edge.*

Fasten two boards together in the checks of the bench screw, at one end, and support the other end with the side pin, inserted in one of the holes of the side board; plane the upper edges as straight as the eye can observe: unscrew the check board, place one board upon the other, with the planed edges together, and the faces of the boards in a straight line with each other; then if the edges coincide, they are straight, but if not, they will be alike round or alike hollow; the prominent parts must be marked, and the operation repeated as often as may be found necessary. In shooting the edges, the rough is first taken off with the jack plane: in convex places, stand still, drawing and pushing the plane to and from you by the motion of the arms, until the prominent part or parts have been reduced by repeated shavings, which will be taken off the wood, every time the plane is driven forwards; then having got the edges very nearly straight, you may take one or two shavings by going the whole length from the hind to the fore end, without drawing back the plane; then with the trying or long plane walk from end to end as before, pushing the plane continually forward, and if it take a shaving of unequal breadth, or unequal thickness, or both, repeat the operation again until this is not the case. If the edges are very long, the same operation must be performed with the jointer, viz. by pushing it forward from end to end. Then, when two edges coincide in working them together in this manner, you will have two straight edges. Straight edges are easier made when the board has been previously faced. Here the workman must keep the definition of a straight line continually in view.

---

§ 73. *To face a Piece of Stuff.*

Here the workman must not lose sight of the definition of a straight surface, viz. it is that which will every where coincide with a straight line : apply the edges of a pair of winding sticks, one to the farther end of the surface, and the other to the nearer; directing the eye* in any straight line coinciding with the upper edges : then if by keeping the eye at the same point, and if straight

---

* That is, shutting one eye and observing with the other. This depends on vision being always performed in straight lines.

lines can be directed from it to all other points in the upper edge of each winding stick, then the ends of the surface are in a plane. Draw a line by the edge of each winding stick on the surface, and if the surface will every where coincide with a straight line, then it is already straight, there will be very little to do but plane the rough away. But if on applying the edges of the winding sticks to the surface, a straight line can only be directed from the eye to one point in the upper edge of each winding stick, then the surface is said to wind, and is called a winding surface; in such a case, there will always be two corners of the surface higher than the other two : then with the jack plane, reduce the surface at the corners, until both edges of the winding sticks are in the same plane; draw a line by the edge of each winding stick on the surface as before, then with the jack plane reduce all the prominent parts between the lines : having obtained a surface very nearly straight by one or several trials by the jack plane, plane off the ridges which the jack plane has left, with the trying plane, and apply the winding sticks in the same manner, in order to be certain whether you are keeping the surface true or not.

----

§ 74. *To shoot the Edge of a Board.*

First rough plane the side of the board with the jack plane, or plane the rough off the side of the board next to the joint.   Then setting the sides of the board in a vertical position, and placing it in the bench screw, proceed in the same manner in the operation of planing, as in making a straight edge; except that there is only one edge planed at a time in shooting.   If the joint is not very long, it is brought to a straight by the eye : but if very long, a straight edge must be used ; in shooting the edge, the hand must be carried regular from end to end.

§ 75. *To join two Boards together.*

Shoot the edge of each board first, or if they are very thin, they may be shot together: apply each of the edges together, then if they are quite close, both face and back of the board, and the faces of the two boards straight with each other, they may be glued together; but if not, the operation must be repeated until there is no space left on either side, and the sides quite straight with each other: when properly shot, spread the edges over with strong thin glue, of a proper consistence, made very hot, one of the boards being fixed, the faces adjoining each other, and the edges straight; then turn the loose board upon the fixed board, applying the edges that are shot together, rub the upper board backwards and forwards until the two begin to stick fast, and the glue mostly rubbed out; the faces must be brought as nearly straight as possible.

----

§ 76. *To join any number of Boards, edge to edge, with glue, so as to form one Board.*

First shoot the edges of two boards, so as to bring them to a joint, mark the faces of these boards next to the joint, then shoot the other edge of one of the boards, and another edge of another board, and bring these to a joint also, marking them as before, proceed in this manner until as many boards have been jointed as make the entire breadth required, always numbering the boards in regular order. Glue the first two together; when sufficiently dry, glue the second and third board, and so on till the joints are glued.

If the boards or planks be very long, the edges which are to be united, will require to be warmed before a fire. And in order to keep the faces fair with each other, three men will be necessary also in helping to rub, one to guide the middle, and one to guide each end.

R

§ 77. *To square and try-up a Piece of Stuff.*

First face the side of the stuff, apply the edge of the stock of a
square to this side, and the edge of the tongue to the other side
or edge to be planed, keeping the stock of the square at right
angles to the arris ; try the square in the same manner in several
places, then plane the side or edge of the stuff, until the inner edge
of the tongue coincide with one side or edge of the stuff, while
the inner edge of the stock coincides with the face.

---

§ 78. *To try-up a Piece of Stuff all round.*

When the two sides of the face and edge have been squared,
gauge the stuff to its thickness by the gauge, then plane the other
side to the gauge line opposite to the face, but observe that it
must be planed so as to coincide with the blade of the square,
while the stock coincides with the other side, on which the gauge
line was drawn, both handle and tongue being at the same time at
right angles to the arris. Having now finished three sides, set
the gauge to the intended breadth, then apply the guide of the
head of the gauge upon the edge or side that is wrought, and
which adjoins the other two wrought sides, and the stem and tooth
upon the side to be gauged, draw a line upon that side, turn the
stuff over to the other side, and place the head upon the same side
as before, but not upon the same edge, and the tooth end of the
stem upon the side of the wood ; draw a line upon this side. In
gauging, you must press the head of the gauge pretty hard against
the surface of the stuff on which it rests, otherwise the grain of
the wood will be liable to draw the tooth of the gauge out of its
straight lined course ; then by working the wood between the
gauge lines straight across, the piece of stuff will be completely
tried-up, and this last side will be planed up without the use of
the square : and, indeed, the third side might also have been done
when the rough edge, whence the gauge line was drawn, is pretty
near the square.

§ 79. *To rebate a Piece of Stuff.*

First, when the rebate is to be made on the arris next to you, the stuff must be first tried-up on two sides; if the rebate is not very large, set the guide of the fence of the moving fillister to be within the distance of the horizontal breadth of the intended rebate; and screw the stop, so that the guide may be something less than the vertical depth of the rebate from the sole of the plane; set the iron so as to be sufficiently rank, and to project equally below the sole of the plane; make the left hand point of the cutting edge flush with the left hand side of the plane: the tooth should be a small matter without the right hand side of the plane. Proceed now to gauge the horizontal and vertical dimensions of the rebate: begin your work at the fore end of the stuff; the plane being placed before you, lay your right hand partly on the top hind end of the plane, your fore fingers upon the left side, and your thumb upon the right, the middle part of the palm of the hand resting upon the round of the plane between the top and the end; lay the thumb of your left hand over the top of the fore end of the plane, bending the thumb downwards upon the right hand side of the plane, while the upper division of the fore-finger, and the one next to it goes obliquely on the left side of the plane, and then bends with the same obliquity to comply with the fore end of the plane; the two remaining fingers are turned inwards: push the plane forward without moving your feet, and a shaving will be discharged equal to the breadth of the rebate; draw the plane towards you again to the place you pushed it from, and repeat the operation: proceed in this manner until you have gone very near the depth of the rebate; move a step backward, and proceed as before; go on by several successive steps, operating at each one as at first, until you get to the end; then you may take a shaving or two the whole length, or take down any protuberant parts.

In holding the fillister, care must be taken to keep the sides vertical, and consequently the sole level: then clean out the bot.

tom and side of the rebate with the skew-faced rebate plane, that is, plane the bottom and side smooth, until you come close to the gauge lines: for this purpose the iron must be set very fine, and equally prominent throughout the breadth of the sole.

If your rebate exceeds in breadth the distance which the guide of the fence can be set from the right side of the plane, you may make a narrow rebate on the side next to you, and set the plow to the full breadth, and the stop of the plow to the depth: make a groove next to the gauge line: then with the firmer chisel, cut off the wood between the groove and the rebate, level with the bottom; or should the rebate be very wide, you may make several intermediate grooves, leaving the wood between every two adjacent grooves of less breadth than the firmer chisel, so as to be easily cut out; having the rebate roughed out, you may make the bottom a little smoother with the paring chisel; then with a common rebate plane, about an inch broad in the sole, plane the side of the bottom next to the vertical side, and with the jack plane take off the irregularities of the wood left by the chisel: smooth the farther side of the bottom of the rebate with the skew rebate plane, as also the vertical side: with the trying plane smooth the remaining part next to you until the rebate is at its full depth. If any thing remain in the internal angle, it may be cut away with a fine set paring chisel; but this will hardly be necessary when the tools are in good order.

When the breadth and depth of the rebate is not greater than the depth which the plow can be set to work, the most expeditious method of making a rebate, is by grooving it within the gauge lines on each side of the arris, and so taking the piece out without the use of the chisel: then proceed to work the bottom and side of the groove as before. By these means you have the several methods of rebating, when the rebate is made on the left edge of the stuff: but if the rebate is formed from the right hand arris, it must be planed on two sides, or on one side, and an edge as before; place the stuff so that the arris of the two planed sides may be next to you. Set the sash fillister to the whole breadth of the stuff that is to be left standing, and the stop to the depth, then you may proceed to rebate as before.

§ 80. *To rebate across the Grain.*

Nail a straight slip across the piece to be rebated, so that the straight edge may fall upon the line which the vertical side of the rebate makes upon the top of the stuff, keeping the breadth of the slip entirely to one side of the rebate ; then having set the stop of the dado grooving plane to the depth of the rebate, holding the plane vertically, run a groove across the wood, repeat the same operation in one or more places in the breadth of the rebate, leaving each interstice or standing-up part something less than the breadth of the firmer chisel: then with that chisel cut away these parts between every two grooves, but be careful in doing this that you do not tear the wood up ; pare the bottom pretty smooth, or after having cut the rough away with the chisel, take a rebating plane with the iron set rather rank, and work the prominent parts down to the aforesaid grooves nearly. Lastly, with a fine set screwed rebating plane, smooth the bottom next to the vertical side of the rebate ; the other parts of the bottom may be taken completely down with a fine set smoothing plane : in this manner you may make a tenon of any breadth.

§ 81. *To frame two Pieces of Stuff together.*

For this purpose it will be necessary to face-up, and square each of the pieces at least on two sides ; the thickness of the tenon or width of the mortise ought not to exceed in general one-third of the thickness of the stuff; but this will in some cases depend upon the work, and whether the materials that are to be framed together be of the same kind or not, and consequently the proportion greater or less according as the piece on which the tenon is cut, is of a stronger or weaker texture than the piece which is to receive it. If the two pieces are to be joined at a right angle, and the piece which has the mortise project only on one side of the piece which has the tenon, you must then set the mortise a little farther in than the breadth of the piece which has the tenon, in order to prevent the piece at the end of the tenon from splitting; mark the length of your tenon a little more than the breadth of the mortised piece; strike a square line through the mark; then at the place where the line meets the arris, strike another square line: if the work is to be very nicely put together, this will be best done with the drawing knife: square two pencil lines on the two sides of the mortised piece opposite to, or in the same straight line with, the inside of the tenoned piece; strike other two square pencil lines upon the sides of the mortised piece next to the end opposite to the outer edge of the tenoned piece, or in the same straight line with it, and thus the distance between each pair of square lines upon each of the sides, will be equal to the breadth of the tenoned piece; but this distance would be too long for the mortise, as when finished, one piece of stuff does not pass by the breadth of the other; therefore if the mortise came close to the end, there would be nothing to resist and keep the tenon in its place: for this reason, the mortise must never be cut out to the extremity, but always at least one fourth of the whole breadth farther in; if the insides of the pieces are intended to be entirely square, you may make the length of the mortise from the inside pencil lines equal to, or nearly two-thirds of the breadth of the tenoned piece. Set the distance of the teeth of the mortise gauge equal to the thickness of the tenon or breadth of the mortise, and the distance from, and of the nearer tooth to the head, equal to the thickness of the cheek of the mortise or shoulder of the tenon, then gauge both pieces on the inner edges from the face, and also on the outer edges from the same face, return the pencil lines upon the outer edge of the mortised piece. Lay the piece to be mortised upon the mortise stool, with the side uppermost which is to be the inside, and mortise half through: turn the other edge uppermost, and mortise the other half: the reason of mortising one half at a time is obvious, when it is considered, that the holding of the mortise chisel at right angles to the surface, is all guess work; the mortise would therefore be liable to go not only obliquely, but uneven : the length of the mortise must be a little more on the outer edge than on the inner, as the tenon when it comes to be stationed to its place, is secured by wedges and glue : the ends of the mortise must be quite straight, though inclining towards each other next to the inside or shoulder of the tenon ; the sides of the cheeks of the mortise must be cut smooth with the paring chisel: and for the purpose of having the width of the mortise, when finished, the exact thickness of the tenon, the mortise chisel ought to be rather of less thickness than that of the tenon.

To form the tenon: cut the shoulders in with the drawing knife; place the side hooks at right angles to the sides of the bench, the knob or catch of each against the side board ; place the tenoned piece upon the side hooks, and against the other knobs on the bench, and with the tenon saw cut the shoulders of the tenon on one side, and turn the other side up and cut the other shoulder ; take the piece and fix it in the bench screw, and with a hand saw cut off the two outside pieces called the tenon cheeks from the sides of the tenon, keeping the stuff entire between the gauge lines ; and if the saw is in good order, it will not be necessary to do any more to the sides; but if the saw has been led away from the draughts, either from carelessness or from its beng in bad order, recourse must be had to the paring chisel, so as to take away the superfluous wood to the gauge lines, and lastly to the skew-faced rebate plane. Having finished the sides of the tenon, it must be reduced from the outer edge to a breadth equal to the length of the mortise, this reduction is called haunching, but it is better to have a little piece to project beyond the shoulder, and then to cut a shallow mortise of the same depth close to the far

ther end of the mortise piece; this little tenon is called stump
haunchings. Insert the tenon in a mortise, driving the end of the
tenoned piece with a mallet, until the shoulder comes home to the
face of the mortise: then if your work has been truly tried-up and
set out, both shoulders will be quite close to the inner edge of the
mortised piece; having thus finished the mortise and tenon, you
may take it out and glue the shoulders of the tenon and inner edge
of the mortise with very hot glue; then drive the tenoned piece
home; if very stiff, it will be necessary to use a cramp; however,
the use of this will be better understood in making a complete
frame.

§ 82. *Boarding Floors.*

Boarded floors are those covered with boards. The operation of
boarding floors should commence as soon as the windows are in,
and the plaster dry. The preparation of the boards for this purpose is as follows:

They should first be planed on their best face, and set out to
season till the natural sap is quite exhausted; they may then be
planed smooth, shot and squared upon one edge: the opposite
edges are brought to a breadth, by drawing a line on the face parallel to the other edge, with a flooring gauge; they are then
gauged to a thickness with a common gauge, and rebated down
on the back to the lines drawn by the gauge.

The next thing to be done is to try the joists, whether they be
level or not: if they are found to be depressed in the middle, they
must be furred up, and if found to protuberant must be reduced
by the adze. The former is more generally the case.

The boards employed in flooring are either battens or deals of
greater breadth. The quality of battens are divided into three
kinds; the best is that free of knots, shakes, sap-wood, or cross-
grained stuff, and well matched, that is, selected with the greatest
care; the second best is that in which only small, but sound knots

are permitted, and free of shakes and sap-wood; the most common kind is that which is left, after taking away the best and second best.

With regard to the joints of flooring boards, they are either quite square, plowed and tongued, rebated, or doweled; in fixing them they are nailed either upon one or both edges; they are always necessarily nailed on both edges, when the joints are plain or square without dowels. When they are doweled, they may be nailed on one or both edges; but in the best doweled work, the outer edge only is nailed, by driving the brad obliquely through that edge without piercing the surface of the board; so that the surface of the floor, when cleaned off, appears without blemish.

In laying boarded floors, the boards are sometimes laid one after another; or otherwise, one is first laid, then the fourth, leaving an interval somewhat less than the breadth of the second and third together: the two intermediate boards are next laid in their places, with one edge upon the edge of the first board, and the other upon that of the fourth board; the two middle edges resting upon each other, and forming a ridge at the joint; to force down these joints, two or more workmen jump upon the ridge till they have brought the under sides of the boards close to the joints, then they are fixed in their places with brads. In this last method, the boards are said to be folded. Though two boards are here mentioned, the most common way is to fold four at a time; this mode is only taken when the boards are not sufficiently seasoned, or suspected not to be so. In order to make close work, it is obvious that the two edges forming the joint of the second and third boards, must form angles with the faces, each less than a right angle. The seventh board is fixed as the fourth, and the fifth and sixth inserted as the second and third, and so on till the completion.

The headings are either square, splayed, or plowed and tongued. When it is necessary to have a heading in the length of the floor, it should always be upon a joist. One heading should never meet another.

Nos. 9 & 10.   s

When floors are doweled, it is better to place dowels over the middle of the interjoist, than over the joists, ·in order to prevent the edge of one board from passing that of the other. When the boards are only braded upon one edge, the brads are most frequently concealed, by driving them slanting through the outer edge of every successive board, without piercing the upper surface. In adzing away the under sides of the boards opposite to the joists, in order to equalize their thickness, the greatest care should be taken to chip them straight, and exactly down to the rebates, as the soundness of the floor depends on this.

---

§ 83. *Hanging of Shutters to be cut.*

Shutters to ne cut must first be hung the whole length, and taken down and cut: but observe that you do not cut the joint by the range of the middle bar, but at right angles to the sides of the sash frame, for unless this be done, the ends will not all coincide when folded together. In order to hang shutters at the first trial, set off the margin from the bead on both sides, then take half the thickness of the knuckle of the hinge, and prick it on each side from the margin, so drawn towards the middle of the window, at the places of the hinges; put in brads at these pricks; then putting the shutter to its place, screw it fast, and when opened it will turn to the place intended.

---

§ 84. *Hanging of Doors.*

Doors should be hung so as to rise above the carpet; for this purpose, the knuckle of the bottom hinge should be made to project the whole pin beyond the surface of the door, while the centre of the upper pin comes rather within the surface. To render this still more effectual, the floor is sometimes raised immediately under the door. A door wider at the bottom than at the top, in a trapezoidal form, will also have the effect of clearing the floor : most of the ancient doors were of this figure.

---

§ 85. *To Scribe one Piece of Board or Stuff to another.*

When the edge end or side of one piece of stuff is fitted close to the superficies of another, the former is said to be scribed to the latter. Thus the skirting boards of a room should be scribed to the floor: In moulded framing, the moulding upon the rails, if not quirked, are scribed to the styles, and muntins upon rails. To scribe the edge of a board against any uneven surface: lay the edge of the board over its place, with the face in the position in which it is to stand : with a pair of stiff compasses opened to the widest part, keeping one leg close to the uneven surface, move or draw the compasses forward, so that the point of the other leg may mark a line on the board, and that the two points may always be in a straight line parallel to the straight line in which the two points were at the commencement of the motion : then cut away the wood between this line and the bottom edge, and the one will coincide with the other.

---

§ 86. *Doors.*

Doors ought to be made of clean good stuff, firmly put together, the mitres or scribings brought together with the greatest exactness, and the whole of their surfaces perfectly smooth ; particularly those made for the best apartments of good houses. In order to effect this, the whole of the work ought to be set out and tried-up with particular care ; saws and all other tools must be in good order ; the mortising, tenoning, plowing, and sticking of the mouldings, ought to be correctly to the gauge lines ; these being strictly attended to, the work will of necessity, when put together, close with certainty ; but if otherwise, the workman must expect a great deal of trouble in paring the different parts before the work

can be made to appear in any degree passable; this will also occasion a want of firmness in the work, particularly if the tenons
and mortises are obliged to be pared.

In bead and flush doors, the best way is to mitre the work
square, afterwards put in the panels, and smooth the whole off
together; then, marking the panels at the parts of the framing they
agree to, take the door to pieces, and work the beads on the
stiles, rails, and muntins.

If the doors are double margin, that is, representing a pair of
folding doors, the staff stile which imitates the meeting stiles, must
be centred to the top and bottom of the door, as well as the hanging; and lock stiles, by forking the ends into notches, cut in the
top and bottom rails.

§ 87. *Stairs.*

Stairs are one of the most important things to be considered in
a building, not only with regard to the situation, but as to the design and execution: the convenience of the building depends on
the situation; and the elegance on the design and execution of the
workmanship. A stair-case ought to be sufficiently lighted, and
the head-way uninterrupted. The half paces and quarter paces
ought to be judiciously distributed. The breadth of the steps ought
never to be more than fifteen inches, nor less than ten; the height
not more than seven, nor less than five; there are cases, however,
which are exceptions to all rule. When you have the height of
the story given in feet, and the height of the step in inches, you
may throw the feet into inches, and divide the height of the story
in inches by the height of the step; if there be no remainder, or
if the remainder be less than the half of the divisor, the quotient
will show the number of steps; but if the remainder be greater
than the half of the divisor, you must take one step more than the
number shown by the quotient: in the two latter cases, you must
divide the height of the story by the number of steps, and the quotient will give the exact height of a step: in the first case, you

have the height of the steps at once, and this is the case whatever description the stairs are of. In order that people may pass
freely, the length of the step ought never to be less than four
feet, though in town houses, for want of room, the going of the
stair is frequently reduced to two feet and a half.

Stairs have several varieties of structure, which depends principally on the situation and destination of the building. Geometrical stairs are those which are supported by one end being fixed
in the wall, and every step in the ascent having an auxiliary
support from that immediately below it, and the lowest step, consequently from the floor.

Bracket stairs are those that have an opening or well, with
strings and newels, and are supported by landings and carriages,
the brackets mitering to the ends of each riser, and fixed to the
string board, which is moulded below like an architrave.

Dog-legged stairs are those which have no opening or well-
hole, the rail and balusters of both the progressive and returning
flights fall in the same vertical planes, the steps being fixed to
strings, newels and carriages, and the ends of the steps of the
inferior kind, terminating only upon the side of the string, without
any housing.

---

§ 88. Of Dog-legged Stairs.

The first thing is to take the dimensions of the stair and height
of the story, and lay down a plan and section upon a floor to the
full size, representing all the newels, strings, and steps: by this,
the situation of string boards, pitching pieces, rough strings, long
bearers, cross bearers, and trimmers will become manifest : the
quantity of room allowed for the stairs, the situation of apertures
and passages, will determine whether there are to be quarter
paces, half paces, one quarter or two quarter winders. In this
description, in order to give all the variety possible, we shall
suppose the flight to consist of two quarter winders.

The strings, rails, and newels being framed together, they must then be fixed, first with temporary supports, the string board will show the situation of the pitching pieces, which must be put up next in order, wedging the one end firmly into the wall, and fixing the other end to the string board; this being done, pitch up the rough strings, and thus finish the carriage part of the flyers. In dog-leg staircases, the steps and risers are seldom glued up, except in cases of returned nosings; we shall therefore suppose them to be separate pieces, and proceed to put up the steps: place the first riser to its situation; having fitted it down so as to be close to the floor, the top being brought to a level at its proper height, and at the same time, the face in its right position, fix it with flat headed nails, driving them obliquely through the bottom part of the riser into the floor, and then nailing the end to the string board. Proceed then to cover the riser with the first tread, observing to notch out the farther bottom angle opposite the rough strings, so as to make it to fit closely down to a level on the top side, while the under side beds firmly upon the rough strings at the back edge, and to the riser towards the front edge; nail down the tread to the rough strings, driving the nails from the seat or place on which the next riser stands, through that edge of the riser into the rough strings, and then nailing the end to the string board; begin with the second riser, having brought it to a breadth, and fitted it close to the top side of the tread, so that the back edge of the tread below it may entirely lap over the back of the riser, while the front side is in its regular vertical position; nail the head to this riser, from the under side, taking care that the nails do not go through the face of the riser, for this would spoil the beauty of the work.

Proceed in this manner as in the last, with tread and riser alternately, until the last parallel riser. The face of this riser must stand the whole projection of the nosing back from the face of the newel. Then fix the top of your first bearer for the first winding tread, on a level with the top of the last parallel riser, so that the farther edge of this bearer may stand about an inch forward from the back of the next succeeding riser, for the purpose of nailing the treads to the risers upwards, as was done in the treads and risers of the flyers, and having fitted the end of this bearer against the back of the riser, and nailed or screwed it fast thereto; this being done, fix a cross bearer, by letting it in half its thickness, into the adjacent sides of the top of the riser, and into the top of the long bearer, so as not to cut through the horizontal breadth of the long bearer, nor through the thickness of the riser, for this would weaken the long bearer, and spoil the look of the riser. Then fix the riser to the newel, driving a nail obliquely from the top edge of the riser into the newel; you may then proceed to put down the first winding tread, fitting it close to the newel, in the bird's-mouth form; proceed with all the succeeding risers and heads, always fixing in the bearers previously to the laying of each successive tread, until the steps round the winding part are entirely completed. Proceed then with the upper retrogressive range of flyers, as those below. Fit the brackets into the backs of the risers and treads, so that their edges may join each other upon the sides of the rough strings to which they are fixed by nails, and thus the work is completed. There are some workmen who do not mind the close fitting to the riser; but certainly it makes the firmest work.

In the best kind of dog-leg stairs, the nosings are returned, and sometimes the risers mitred to brackets, and sometimes mitred with quaker strings: in this case, there is a hollow mitred round the internal angle of the under side of the tread, and the face of the riser. Sometimes the string is framed into the newel, and notched to receive the ends of the steps, and at the other end a corresponding notch board, then the whole flyers are put up as a step ladder.

In order to get the lower part for the turning, set on the thickness of the capping on the return string board, and where that falls on the newel below, is the place of the under limit of the turning.

To find the section of the cap of the newel for the turner, draw a circle to its intended diameter, draw a straight line from the centre to any point without the circumference, and set half the breadth of the rail on each side of that line, and through the point draw a line parallel to the middle straight line, then the two extreme lines will contain the breadth of the rail: draw any radius of the circle, and set half the breadth of the rail from the centre toward the circumference, and through the point where this breadth falls, draw a concentric circle from the point where this circle cuts the middle line of the rail; draw two lines to the points where the breadth of the rail intersects the outer circle, and these lines will show the mitre. The section may then be found by the following method.

After having drawn the outline of the cap and rail as above, take a small portion of the rail, and cut it to the mitre as drawn, then take a block of sufficient size for the cap, and cut out the internal mitre of the cap to answer the external mitre of the rail: place the mitre of the rail into its mitre socket, and draw a line where the surface of the piece meets the mitre; draw the middle line of the rail upon both sides of the block, which will bisect each mitre; take the distance from the centre of the circle above drawn to the mitre point, and set it on each side of the block for the cap upon the middle line of the breadth of the rail, from the mitre point towards the centre of the block, pricking the block at the other extremity of this distance; then these points will be the centres for turning. Fit a piece of wood to the internal mitre, pare off the top part of this piece next to the mitre of the cap, so as to correspond to the line drawn by the top of the rail, then with weak glue stick in this piece to its birth; and being so fitted, send it to the turner.

In order to eradicate a prevalent false idea which many workmen entertain, when the outer edge of the mitre cap is turned so as to have the same section as that of the rail, they suppose this to be all that is necessary for the mitring of the above: but from a very little investigation of the nature of the lines, they will easily be convinced that the sides of the mitre can never be straight surfaces cr planes, but must be curved, when this is the case.

§ 89. *Bracket Stairs.*

The sam methods must be observed with regard to taking the dimensions, and laying down the plan and section, as in dog-legged stairs. In all stairs whatever, after having ascertained the number of steps, take a rod the height of the story, from the surface of the lower floor to the surface of the upper floor: divide the rod into as many equal parts as there are to be risers, then if you have a level surface to work upon below the stair, try each one of the risers as you go on; this will prevent any excess or defect, which even the smallest difference will occasion: for any error, however small, when multiplied, becomes of considerable magnitude, and even the difference of an inch in the last riser, being too high or too low, will not only have a bad effect to the eye, but will be apt to confound persons not thinking of any such irregularity. In order to try the steps properly, by the story rod, if you have not a level surface to work from, the better way will be to lay two rods or boards, and level their top surface to that of the floor, one of these rods being placed a little within the string, and the other near or close to the wall, so as to be at right angles to the starting line of the first riser, or, which is the same thing, parallel to the plan of the string; set off the breadth of the steps upon these rods, and number the risers; you may set not only the breadth of the flyers, but that of the winders also. In order to try the story rod exactly to its vertical situation, mark the same distances on the backs of the risers upon the top edges, as the distances of the plan of the string board and the rods are from each other.

As the internal angle of the steps is open to the end, and not closed by the string, as in common dog-legged stairs, and the neatness of workmanship is as much regarded as in geometrical stairs, the balusters must be neatly dove-tailed into the ends of the steps two in every step; the face of each front baluster must be in a straight surface with the face of the riser; and as all the balusters must be equally divided, the face of the middle baluster must in course stand in the middle of the face of the riser of the preceding

step, and the face of the riser of the succeeding step. The risers
and treads are all glued and blocked previously together; and when
put up, the under side of the step nailed or screwed into the under
edge of the riser, and then rough bracked to the rough strings, as
in the dog-legged stairs; the pitching pieces and rough strings being
similar to those. In gluing up the steps, the best method is to make
a templet, so as to fit the external angle of the steps with the nosing.

---

§ 90. *Geometrical Stairs.*

The steps of geometrical stairs ought to be constructed so as to
have a very light and clean appearance when put up: for this
purpose, and to aid the principle of strength, the risers and treads
when planed up, ought not to be less than one inch and an eighth,
supposing the going of the stair or length of the step to be four
feet; and for every six inches in length, you may add one-eighth
part more; the risers ought to be dove-tailed into the cover, and
when the steps are put up, the treads are screwed up from below,
to the under edges of the risers; the holes for sinking the heads
of the screws ought to be bored with a centre bit, and then fitted
closely in with wood well matched, so as to conceal the screws
entirely, and to appear as one uniform surface without blemish.
Brackets are mitred to the riser, and the nosings are continued
round: in this mode, however, there is an apparent defect, for the
brackets, instead of giving support, are themselves unsupported,
depending on the steps, and are of no other use in point of
strength, than merely tying the risers and treads of the internal
angles of the steps together; and from the internal angles being
hollow, or a re-enterant right angle, except at the ends, which
terminate by the wall at one extremity, and by the brackets at the
other, there is a want of regular finish. The cavetto or hollow
is carried all round the front of the slip returned at the end, returned again at the end of the bracket, thence along the inside of
the same, and then along the internal angle of the back of the riser.

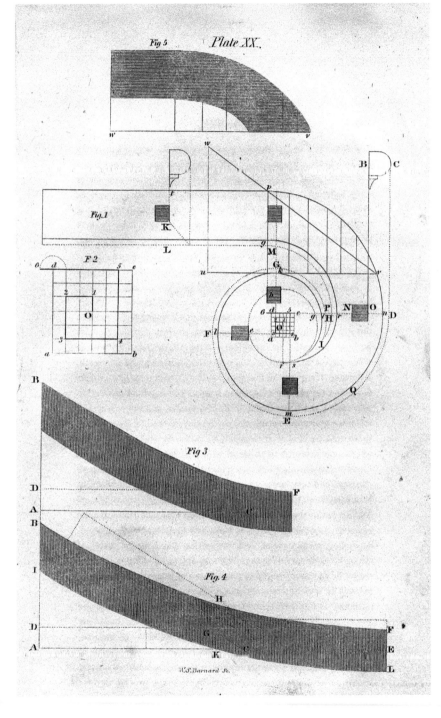

Fig 5

Plate XX.

Fig.1

F 2

Fig 3

Fig. 4

W.S.Barnard Sc.

This is a slight imitation of the ancient mode, which was to make the steps solid all the way, so as to have every where throughout its length a bracket-formed section. This, though more natural in appearance, would be expensive and troublesome to execute, particularly when winders are used, but much stronger.

The best mode, however, of constructing geometrical stairs, is to put up the strings, and to mitre the brackets to the risers as usual, and finish the soffit with lath and plaster, which will form an inclined plane under each flight, and a winding surface under the winders. In elegant buildings, the soffit may be divided into panels. If the risers are got out of two inch stuff, it will greatly add to the solidity. The following is the method of drawing and executing the scroll and other wreathed parts of the hand rail.

PLATE XX.

To describe the scrool of a Hand Rail.

From any convenient point, *o, Fig.* 1, *Plate* 20, as a centre describe a circle, *e f g h,* and describe the square *a b c d,* of which the centre is the same, point o, and of which the sides *a b, b c, c d, d a,* are each one third of the diameter of the circle. Divide each side of the square into six equal parts, (see also *Fig.* 2, drawn at large,) and through the points of division draw lines parallel to the sides; taking the distance 0, 1, equal to the side of one of the lesser squares; the distance from 1 to 2, equal in length to twice the side of one of the little squares; the distance from 2 to 3, equal to three times the side of one of the little squares, and so on increasing each line by the side of one of the lesser squares, for the other distances, from 3 to 4, and from 4 to 5, and from 5 to 6, in such a manner, that the distance from 1 to 2, from 2 to 3, from 3 to 4, from 4 to 5, from 5 to 6, may be respectively perpendicular to the distances from 0 to 1, from 1 to

2, from 2 to 3, from 3 to 4, and from 4 to 5. See the centres constructed to a larger scale, *Fig.* 2.

Let the distance between 1 and 0 be produced to meet the circle f. Then from the centre 1, with the radius if, describe the quadrant $f\,i$; from the centre 2, with the distance 2 i, describe the quadrant $i\,k$; from 3, with the distance 3 k, describe the quadrant $k\,l$; from 4, with the distance 4 l, describe the quadrant 1 m; from 5, with the distance 5 m, describe the quadrant m n, and lastly from 6, with the radius 6 n, describe the quadrant n p.

In the radius 6 p, make p q equal to the breadth of the hand-rail, say two inches; then, from the centre 6, describe the quadrant q r, meeting the radius 5 n in r; from the centre 5, with the radius 5 r, describe the arc r s, which will complete the scroll. The shank of the scroll is drawn from the points p and q, parallel to the radius 6 n.

Curtail step is the lowest step of the stair, one end being formed into a spiral, agreeable to the scroll of the hand rail.

Let the balusters be so placed, that the distances between every two nearest may be all equal, that two balusters may stand upon every step, and that the front of a baluster may be in the plane of every riser. Likewise that the middle of every baluster, may be in the middle of the breadth of the hand-rail, and that the middle of the balusters which support the scroll, may also be in the middle of the breadth of the scroll, and that the outer edges of these balusters, may be all at the same distance from the curved edge of the curtail step, and lastly, that none of the distances between those which support the scroll, be greater than those which rise from the flyers to support the rail, and that the distances of the balusters which rise from the curtail step to support the scroll, may be less and less, as the radius of curvature for describing the plan of the scroll is less.

Hence the plan of the rail will be in the middle of the plan of the curtail step, and because the projection of the nosing from the outside the baluster, is greater than the projection of the outside of the rail, the breadth of the scroll of the hand-rail, will be less than the breadth of the scroll worked on the end of the curtail step.

To describe the Curtail Step.

Let the balusters be properly placed in the order now described, and let k be the angle of the baluster in the front of the second riser, and let N o be the breadth of a baluster in the breadth of the rail. Make o d each equal to the projection of the nosing, and through the point d describe the spiral D E F G I, to follow the nearest spiral in the place of the rail; also through the point P describe the spiral P M L, so as to be every where equidistant from that which forms the other edge of the plan of the scroll. These two last described spirals may be drawn from the same centres, as those from which the edges of the scroll were drawn. To render this description evident to workmen, B c is a section of the nosing of the front of the curtail step, and c d is a plan of the nosing attached to the straight part of the curtail step. Likewise L M is the plan of the nosing which projects from the strong board.

To describe the Face Mould for the wreathed part of the Scroll.

Draw the straight line u v, parallel to the shank of the scroll, to touch the interior part nearest the centre, and let it meet the outside of the scroll in v. Make v u equal to the breadth of one of the flyers, and draw u w, perpendicular to v u, and make w w equal to the height of a step, and join v w. From the curved edge of the scroll, draw ordinates perpendicular to u v. Transfer the several distances from w, where the ordinate or ordinates produced, intersect v w, to the line v w, *Fig.* 5, and draw the ordinates in *Fig.* 5, and make them respectively equal to the ordinates both of the outside and inside curves of *Fig.* 5, and the ends, then the shadowed figure of *Fig.* 5, will be the face mould.

To find the Falling Moulds.

Let ꞯ be that point in the outside spiral which separates the wreathed part of the scroll from the level next to the centre.

In *Fig.* 4, describe the right-angled triangle A C B, identical to the right-angled triangle *u v w, Fig.* 1, A B being equal to the height of a step. Produce A C to B, and make D E equal to the developement of the next line, *t p v n* ꞯ.

Divide A B into six equal parts, and let D be the first point of division. Draw D F parallel to A E, and E F, parallel to A D, intersecting B C in G. Describe a parabola B H F, to touch G F and G B at B and F; this may be conveniently formed by the intersections of lines; also draw another curve, I K L, under the other, and every where 2 inches distant from it, which distance must be the depth of the rail, and the falling mould, *Fig.* 4, for the convex side will be complete. *Fig.* 3, inside falling mould.

PLATE XXI.

EXPLANATION.

Showing the Construction of a Dog-leg Staircase.

No. 1. the plan.

No. 2. the elevation.

A B, No. 2. the lower newel, the part B C being turned

a No. 1. the seat of the newel on the plan.

G H, No. 2. the upper newel.

g, No. 1. its seat on the plan.

D E and F G, No. 2. lower and upper string boards framed into the newels.

K L, No. 2 a joist framed into the trimmer I.

k l, n o, q r, &c. No. 2. the faces of the risers; *m n, p q, s t,* the treads of the cover boards.

m, p, s, &c. No. 2. the nosings of steps.

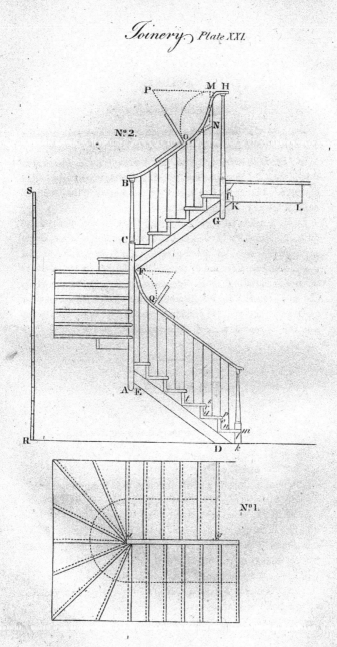

Joinery, Plate XXI.

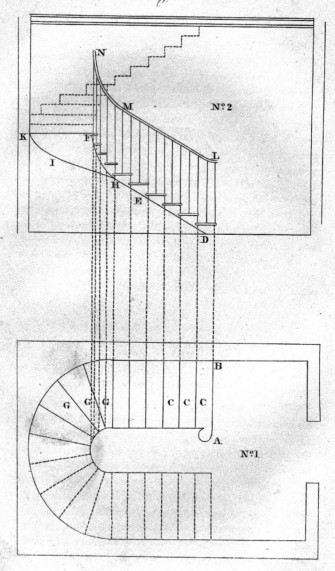

N? 2

N?1

The dotted lines on the plan, represent the faces of the risers; and the continued lines, the nosings of the steps.

M O and F Q, upper and lower ramps.

The method of drawing the ramp is as follows: suppose the upper ramp' to be drawn; produce the top H M of the rail to P draw M N perpondicular to the horizon, and produce the straigh part O N of the pitch of the rail, to meet it in N; making N O equal to N M: draw O P at right angle to O N: from P as a centre, describe the arc M O, and then the other concentric circle, which will complete the ramp required.

R S the story rod; a necessary article in fixing the steps, for if put up only by a common measuring rule, will frequently occasion an excess or defect in the height, so as to render the stair extremely faulty, which cannot be the case, if the story rod is applied to every riser, and the riser regulated thereby. In the aforesaid case, the error is liable to multiply.

———

PLATE XXII.

To draw the Scroll of a Hand-rail.

In Fig. A, make a circle for the eye three inches and a half diameter, divide the diameter into three equal parts, and make a square in the centre to one of those parts, and divide each side of the square into six equal parts; this square is shown in E, at the bottom, in full size for practice, and laid in the same position as the little square above, so that the centres may be more readily found. These centres are all marked in a regular position; the centre at 1 draws from i round to k, the centre at 2 draws from k to l, and the centre at 3 draws from l to m, &c. which will complete the outside revolution at a, with the centre c; then set the thickness of the rail from a to f, and go the reverse way to draw the inside; then the scroll will be completed.

Note. In order to prevent confusion, the letters are not marked in the small square, Fig. **A.**

To draw the Curtail Step.

Set the balusters in their proper places on each quarter of the scroll in Fig. A ; the first baluster shows the return of the nosing round the step, the second baluster is placed at the beginning of the twist, and the third baluster a quarter distant, and straight with the front of the last riser : then the projection of the nosing is set without, and drawn concentric with the scroll, which will give the form of the curtail.

To draw the Face Mould for squaring the Twist Part of the Scroll.

The reader will observe here, that the joint is made at 3, 6, just to clear the side of the scroll ; draw ordinates across the scroll at discretion, to cut the line *d b, a b c* being the pitch board ; take notice that lines be drawn from 3 and 6 to meet *d b,* so that you may have the said points exact at 3 and 6 in your face mould ; then take the line *d b,* and mark the places of the ordinates upon a rod, and transfer the divisions to *d b* in B, then trace B, from Fig. A, according to the letters.

To find the falling Mould C.

In C, *a b c* is the pitch-board ; the height is divided into six parts, to give the level of the scroll ; the distance *a d* is from the face of the riser to the beginning of the twist ; and the distance from *d* to *k* in C, is the stretch-out from *a,* the beginning of the twist to *h* in Fig. A ; each being any point taken at discretion more than the first quarter; divide the level of the scroll, and the rake of the pitch-board, into a like number of parts, and complete the top edge of the mould by intersecting lines, and the under edge parallel to it to the depth of the rail.

To find the parallel thickness of Stuff for the Twist and Scroll.

Extend the curve *a b c d e*, to 6, in Fig. A, upon the base of the pitch-board from *d* to *g*, in Fig. C; draw *g h* perpendicular to intersect with the top of the mould; draw the dotted line *h f*, parallel to the level of the scroll both ways; apply the distance 6 1, in Fig. A, that is, the length of the plan, for the twist part, from *d* to *e* in C, and draw *e f* perpendicular, to cut the parallel *f h*; draw a dotted line through *f*, parallel to *c b*, the longest side of the pitch-board, which gives the thickness of stuff for the twist, about three inches and a half; and the parallel line from *f* to the base, shows the thickness of the scroll.

Note. The falling mould D, for the outside, is found in the same manner as the other falling mould C.

In order to get a true idea of the twist of the hand rail, the section of the rail by a plane passing through the axis of the well-hole or cylinder is every where a rectangle, that is, the plumb or vertical section, tending to the centre of the stair. This rectangle is every where of an equal breadth, but not of an equal vertical dimension in every part of the rail, unless that the risers and treads were every where the same, from the top to the bottom: the height is greatest above the winders, because the tread is of less breadth, and it is of less height above the flyers; the tread being the greatest. If you cut the rail after squaring it, perpendicular to any of its curved sides, the section will not then be a rectangle, three of the sides will at least be curved. Hence two falling moulds laid down in the usual way, will not square the rail, though in wide openings, they may do it sufficiently near

Nos. 11 & 12. ᴛ

Hence in squaring the rail, the square can never be applied at right angles to any one of the four arrises, for the edge of the stock will not comply with the side of the rail, being curved; this would be easily made to appear by making a wreathed part of a rail, of unusual dimensions, and cutting it in both directions. Therefore, to apply the square right, keep the stock to the plumb of the stair; and to guide the blade properly, the stock ought to be very thick, and made concave to the plan, so as to prevent the possibility of its wabbling or turning from side to side; as a little matter up or a little down, in the direction of the blade, would make a great difference in the rectangling or squaring of the rail.

All this might easily be conceived from the cylinder itself, for there is no direction in which a straight line can be drawn on the surface of a cylinder, but one, and this line is in a plane passing through the axis of the cylinder; and as the two vertical surfaces of the rail are portions of cylinders, there can be no straight line upon such surface, but what must be vertical; all others from this principle are curves; or the sections of the rail are bounded by curves, or by a curve on that side.

In gluing a rail up in thicknesses, it will be sufficiently near to get out a piece of wood to the twisted form by two falling moulds, provided the well-hole be not less than one foot diameter; the thickness of this piece, as is there stated, must be equal to the thickness, or rather the horizontal breadth of the rail, together with the thickness which the number of saw kerfs will amount to; and also the amount of the substance, taken away by planing the veneers. We are now supposing the plan of the rail to be semi-circular, with two straight parts, one above and one below; a plan more frequently ado· from motives of economy, than from any propriety of elegance.

The first thing to be done is to make a cylinder of plank to the size of the well-hole. Draw two level lines round the surface of this cylinder, at the top and bottom; upon each of these lines, set off the treads of the steps, at the end next the well-hole. Draw lines between every two corresponding points at the head and foot, and these lines will be all parallel to the axis of the cylinder. Upon the two lines where the cylindric part begins to commence, and also upon a middle line between these lines, set the heights of the winders, and the height of one of the flyers above and below, or as much as is intended to be taken off the straight of the rail. Take a pliable slip of wood, straight on one edge, and bend it round, and keep the straight edge of it upon the three corresponding points at the height of the last rise of the flyer; then draw the tread of the first winding step by the straight edge from the line where the cylindric part commences to the first perpendicular line on the curved surface; take the next three points higher, and draw a line between the second and third perpendicular lines; proceed in like manner with the next three higher points, and draw a line between the next two adjoining cylindric lines, and the lines so drawn between each three points will be the section of the treads of the succeeding winding steps.

Having thus gone through the cylindric part, draw a step at the top, and another at the bottom, and thus the sections of the steps will be completed; draw the hypothenusal or pitch lines of the flyer on the lower part, and that of the upper part, and whatever difference you make in the height of the rail between the flyers and the winders, you must set it up from the nosings of the steps of the winders upon two of the perpendicular lines: draw a line through the two points, by bending a straight edged slip round the cylinder, the straight edge of the slip coinciding with these points; this line will represent the top of the rail over the winders, and the hypothenusal lines at the bottom and top that of the flyers, then curve off the angles at the top and bottom where the rail of the winding parts meets that of the flyers above and below, then a line being drawn parallel to this, will form the falling mould. The reason of making the vertical elevation of the rail more upon the winders than the flyers, is, that the sudden elevation of the winders diminishes the height of the rail in a direction perpendicular to the raking line, and by this means persons would be liable to fall over it.

To lay the veneers upon the cylinder, if bed screws or wedges are used, you may try the veneers first upon the cylinder, screwing them down without glue; prepare several pieces of wood, to lie from six to twelve inches apart, according to the diameter of the well-hole, with two holes in each, distant in the clear something more than the breadth of the rail. Then having marked the positions of the places of these pieces on the cylinder, pierce the cylinder with corresponding holes on each side of the depth of the rail. If the cylinder is made of plank two inches thick, it will be sufficient for the screws : but if of thinner stuff, it will be convenient to set it an end upon stools to get underneath, confining the top with nuts. Unscrew one half, three men being at work, one holding up all the veneers, another gluing, and the third laying them down successively, one after the other, until all are glued: screw them down immediately. Unscrew the other half, and proceed in like manner, and the rail will be glued up. The glue that is used for this purpose, ought to be clear, and as hot as possible: the rail ought likewise to be made hot, as otherwise the glue will be liable to set before all the veneers are put down, and ready for the screws : this operation should therefore be done before a large fire ; the veneers thoroughly heated previous to the commencement, in order that the heat may be as uniformly retained as possible, throughout the process. The glue in the joints of the rail, will take about three weeks to harden in dry weather.

that the light and shade of the adjoining hollows are more contrasted, the angle of their meeting being more acute, than if a flat space were formed between them. See Figs. 12 and 13, fluting round the convex surface of a cylinder

EXPLANATION OF PLATE XXII.

Showing the Construction of Geometrical Stairs.

No. 1. the plan.

No. 2. the elevation or section.

A B, No. 1. the curtail step, which must be first fixed.

C, C, C, &c. flyers supported below upon rough carriages and partly from the string board D H E F, No. 2 ; sometimes the ends next to the wall are housed into a notch board, and the steps made of thick wood, and no carriages used.

G, G, G, &c. winders fixed to bearers, cross bearers, and pitching pieces, when the flyers are supported upon carriages: sometimes the winders are made of strong stuff, firmly wedged into the wall, the steps screwed together, and the other ends of the steps fixed to the string D E H F. The strength of the stair may be powerfully assisted by a bar of wrought iron, made to coincide with the inside, and screwed to the string immediately below the steps ; this would make a very light stair, and if well attended to in the workmanship, will be equal in firmness to one of stone.

H I K, the wall line of the soffit of the stair for winding the part.

L M N, part of the rail supported by two balusters upon every step.

INDEX

EXPLANATION OF TERMS

USED IN

JOINERY.

N B *This Mark § refers to the preceding Sections according to the Number*

———◆———

A.

ARRIS, the line of concourse or meeting of two surfaces.

B.

BARS for sashes, § 70. *See Plate* XIX. *Figs.* 1, 2, 3, 4, 5, 6, 7, 8.

BASIL, § 5.

BATTEN, a scantling of stuff from two inches to seven inches in breadth, and from half an inch to one inch and a half thick, § 82.

BEADS, § 31, 68, 69. *See Plate* XIV. *Figs.* 2, 3. *Plate* XV. *Figs.* 1, 2, 3, 4.

BEAKING JOINT is the joint formed by the meeting of several heading joists in one continued line, which is sometimes the case in folded floors.

BENCH, § 2, 67. *See Plate* XII. *Fig.* 12.

BENCH HOOK, § 2.

BENCH PLANES, § 14. *See Plate* XII. *Figs.* 1, 2, 3.

Bench Screw, § 2.

Bevel. One side is said to be bevelled with respect to another, when the angle formed by these two sides, is greater or less than a right angle.

Bevel, the tool, § 58, 67. *See Plate* XIII. *Fig.* 12.

Bits, § 34. *See Plate* XIII. *Fig.* 1.

Blade is expressed of any part of a tool that is broad and thin, as the blade of an axe, of an adze, of a chisel, of a square. The blade of a saw is more frequently called the plate.

Boarding Floors, § 82.

Bottom Rail, the lowest rail of a door.

Bead, a small nail without any projecting head, except on one edge. The intention is to drive it within the surface of the wood, by means of a hammer and punch, and fill the cavity flush to the surface with putty.

Brad Awl, § 39, 67. *See Plate* XIII. *Fig.* 3.

Brace and Bits, the same as stock and bits.

Breaking Joint, is, not to allow two joints to come together.

C.

Casting or Warping, is the bending of the surfaces of a piece of wood from their original position, either by the weight of the wood, or by an unequal exposure to the weather, or by unequal exture of the wood.

Cavetto, § 68.

Centre Bits, § 35.

Chisels, § 40. *See Plate* XIII. *Figs.* 3, 4, 5.

Cima-Recta, § 68. *See Plate* XIV. *Figs.* 10, 11.

Cima-Reversa, § 68. *See Plate* XIV. *Fig.* 12.

Clamp, a piece of wood fixed to the end of a board, by mortise and tenon, or by groove and tongue, so that the fibres of the one piece thus fixed, transverse those of the board, and by this means prevents it from casting; the piece at the end is called a clamp, and the board is said to be clamped.

CLEAR STORY WINDOWS are those that have no transom.

COUNTERSINKS, § 36.

COMPASS PLANE, § 15.

COMPASS SAW, § 53. *See Plate* XIII. *Fig.* 9.

CROSS-GRAINED STUFF, is wood having its fibres running in contrary positions to the surfaces, and consequently cannot be made
perfectly smooth, when planed in one direction, without turning
it or turning the plane. This most frequently arises from a
twisted disposition of the fibres.   mmmmm   mmmm

CURLING STUFF, is that which is occasioned by the winding or
coiling of the fibres round the boughs of the tree, when they
begin to shoot out of the trunk. The double iron planes now
in use, are a most complete remedy against cross-grained and
curling stuff; the plane will nearly work as smooth against the
grain as with it.

D.

DADO GROOVING PLANES, § 29.

DOOR FRAME, the surrounding case into, and out of which the
door shuts and opens, consisting of two upright pieces and a
head, generally fixed together by mortise and tenon, and
wrought, rebated, and beaded.

DOORS, § 70. *See Plates* XVI, XVII, XVIII.

DOOR HUNG, § 84.

DOUBLE TORUS, § 69. *See Plate* XV

DOVE-TAIL SAW, § 52.

DRAGING in the hanging of doors, is a depression or lowering of
the door, so as to make it rub on the floor, occasioned by the
loosening of the hinges, or the settling of the building.

DRAW BORE PINS, two iron pins with wooden handles, for the purpose of forcing the shoulders of tenons against the abutments on
the cheeks of the mortises, so as make a close joint. Draw
bore pins are in joinery, what hook pins are in carpentry, and
used in a similar manner. *See Carpentry*, § 20.

DRAWING KNIFE, § 44.

E.

EDGE TOOLS, all tools made sharp so as to cut.

F.

FENCE, the guard of a plane which obliges it to work to a certain horizontal breadth from the arris. All moulding planes, except hollows and rounds, and snipesbills, have fixed fences as well as fixed stops, but in fillisters and plows, the fences are moveable, § 20, 21, 22, 23, 28, 31.

FINE SET, when the iron has a very small projection below the sole of the plane, so as to take a very thin broad shaving, it is said to be fine set.

FIRMER CHISEL, § 67. *See Plate* XIII.

FLOORS, § 82.

FORKSTAFF PLANE, § 16.

FRAMING, § 81.

FREE STUFF, that which is quite clean or without knots, and works easily, without tearing.

FROWY STUFF, the same as free stuff.

G.

GAUGE, § 59, 67. *See Plate* XIII. *Fig.* 13.

GIMLET, § 67. *See Plate* XIII. *Fig.* 2. *No.* 1, 2.

GOUGE, § 43.

GRIND STONE, a cylindric stone, which being turned round its axis, edge tools are sharpened by applying the basil to the convex surface.

GRINDING THE IRON, § 6.

GROOVE, § 28.

GROOVING PLANES, § 28. *See Plate* XII. *Fig.* 8, 9. § 2.

H.

HAMMER, *See Carpentry.* § 15.

X

HAND SAW, § 48, 67. *See Plate* XIII. *Fig.* 6.
HANGING DOORS, § 84.
HANGING SHUTTERS, § 83.
HATCHET, § 55.
HINGING DOORS AND SHUTTERS, § 83, 84.
HOLLOWS AND ROUNDS, § 33.

J.

JACK PLANE, § 5, 8, 67. *See Plate* XII. *Fig.* 1.
JOINTER, § 12.

K.

KERF, the way which the saw makes in dividing a piece of wood into two parts.
KEY-HOLE SAW, § 54, 67. *See Plate* XIII. *Fig.* 10.
KNOT, that part of a branch of a tree where it issues out of the trunk.

L

LONG PLANE, § 11.
LOWER RAIL, the rail at the foot of a door next to the floor.
LYING PANEL, a panel with the fibres of the wood disposed horizontally. Lying panels have their horizontal dimensions generally greater than the vertical dimension.

M.

MALLET. *See Carpentry,* § 16. *Joinery,* § 67. *and Plate* XII. *Fig.* 9.
MARGINS or MARGENTS, the flat part of the stiles and rails of framed work.
MIDDLE RAIL, the rail of a door which is upon a level with the hand when hanging freely and bending the joint of the wrest. The lock of the door is generally fixed in this rail.   mmm
MITRE. When two pieces of wood are formed to equal angles, or each two sides of each piece at equal inclinations, and two sides, one of each piece, joined together at their common vertex,

so as to make an angle, or an inclination double to that of either piece, they are said to be mitred together, and the joint is called the mitre. The angle which is thus formed by the junction of the two, is generally a right angle.

MITRE SQUARE, § 66.

MORTISE CHISELS, § 42, 67. *See Plate* XIII. *Fig.* 5.

MORTISE AND TENON, § 81.

MORTISE GAUGE, § 60.

MOULDING PLANES, § 30.

MOULDINGS, § 68, 69, 70. *See Plate* XIV, XV, XVI, XVII, XVIII, XIX.

MOVING FILLISTER, § 20.

MULLION, the large bars or divisions of windows.

MUNNION, a large vertical bar of a window frame, separating two casements or glass frames from each other.

MUNTINS or MONTANTS, the vertical pieces of the frame of a door between the stiles.

O.

OGEE, a moulding, the transverse section of which consists of two curves of contrary flexture, § 68. *See Plate* XIV. *Figs.* 10, 11, 12.

P.

PANEL, a thin board, having all its edges inserted in the grooves of a surrounding frame.

PANEL SAW, § 49.

PLOW, § 28, 67. *See Plate* I. *Fig.* 8.

Q.

QUARTER ROUND, § 68. *See Fig.* 7.

R.

RAILS, the horizontal pieces which contain the tenons in a piece of framing, in which the upper and lower edges of the panels are inserted.

RAISERS. *See Risers.*

Rank Set, is when the edge of the iron projects considerably below the sole of the plane, so as to take a thick shaving.

Rebate, § 18.

Rebating, § 79, 80.

Rebating Planes, § 18, 19, 20, 21, 22, 23, 24, 25, 26, 27, also 67. *See Plate* XII. *Fig.* 6, 7.

Reeded Mouldings, § 69. *See Plate* XV. *Figs.* 7, 8, 9.

Return. In any body with two surfaces joining each other at an angle, one of the surfaces is said to return in respect of the other; or if standing before one surface, so that the eye may be in a straight line with the other, or nearly so; this last is said to return

Rimers, § 37.

Ripping Saw, § 46.

Risers, the vertical sides of the steps of stairs.

Rub Stone, § 6.

S.

Sash Fillisters, § 21, 22. *See Plate* XII. *Fig.* 8.

Sash Saw, § 51, 67. *See Plate* XIII. *Fig.* 8.

Saws, § 45.

Scantling the transverse dimensions of a piece of timber, sometimes also the small timbers in roofing and flooring, are called scantlings.

Scotia, § 68. *See Plate* XIV. *Fig.* 9.

Scribe, § 85.

Shoot, a joint, § 74.

Shooting Block, § 63.

Shutters Hung, § 83.

Side Hook, § 61, 67. *See Plate* XII. *Fig.* 11.

Side Rebating Planes, § 27.

Side Snipesbills, § 32.

Single Torus, § 69. *See Plate* XIV. *Fig.* 5. *Plate* XV. *Fig.* 5

Smoothing Plane, § 13, 67. *See Plate* XII. *Fig.* 3.

Snipesbills, § 32.

Square, § 56, 67. *See Plate* XIII. *Fig.* 11.

Staff, a piece of wood fixed to the external angle of the two upright sides of a wall for floating the plaster to, and for defending the angle against accidents.

STILES of a door, are the vertical parts of the framing at the edges of the door.

STOCK AND BITS, § 34, 67. *See Plate* 13. *Fig.* 1.

STRAIGHT BLOCK, § 17.

STRAIGHT EDGE, § 64.

STUFF, § 1.

SURBASE, the upper base of a room, or rather the cornice of the pedestal of the room which serves to finish the dado, and to secure the plaster against accidents, as might happen by the backs of chairs or other furniture on the same level

T.

TANG OF AN IRON is the narrow part of it which passes through the mortise in the stock.

TAPER, the form of a piece of wood which arises from one end of a piece being narrower than the other.

TENON SAW, § 50, 67. *See Plate* XIII. *Fig.* 7.

TOOTH, a small piece of steel with cutting edge in fillisters and gauges

TORUS, § 69. *See Plate* XIV. *Fig.* 5. *Plate* XV. *Figs.* 5, 6.

TRANSOM WINDOWS, those which have horizontal mullions.

TRUSSELS, four-legged stools for ripping and cross-cutting timber upon. For this purpose there are generally two required, and when the timber is very long, an additional trussel in the middle will be found necessary.

TRY, § 78.

TRYING, § 78.

TRYING PLANE, § 9, 10, 67. *See Plate* XII. *Fig.* 10.

TURNING SAW, § 54, 67. *See Plate* XII[r] *Fig.* 20.

W.

WARP. *See Cast.*

WEB OF AN IRON is the broad part of it which comes to the sole of the plane, the upper edge or end of the web has generally one shoulder, and sometimes two, where it joins the tang.

WINDING STICKS. § 64.

BRICKLAYING.

§1. BRICKLAYING is an art by which bricks are joined and cemented, so as to adhere as one body.

This art, in London, includes the business of walling, tiling,
and paving, with bricks or tiles; and sometimes the bricklayer
undertakes the business of plastering also: but this is only done
by masters in a small way. In the country, bricklaying and plastering are generally joined: and not unfrequently the art of masonry also; which has a nearer affinity to it than that of plastering.

The bricklayer is supplied with bricks and mortar at his work
by a man, called a labourer, who also makes the mortar.

The materials used are mortar, bricks, tiles, laths, nails, and
tile pins; bricks and tiles are of several kinds, which, as well as
other descriptions of work, are treated of under their respective
heads: viz. 1. The tools. 2. Of cements. 3. Of brick-making,
and the various sorts of bricks. 4. The several kinds of tiles and
laths. 5. The different methods of treating foundations, according
to the quality of the soil, whether of an uniform or mixed texture.
6. Walling. 7. A description of the plates. Lastly, An explanation of such terms as have not been defined in the course of the
work, or such as may require a farther explanation ; with an index
to the principal technical terms used in this art, and in connection
therewith, the terms and index being placed under an alphabetical
arrangement, as to the former branches of carpentry and joinery.

BRICKLAYING.

BRICKLAYING TOOLS DESCRIBED.

§ 2. *A List of Walling Tools.*

1. A brick trowel , 2. a hammer; 3. a plumb rule; 4. a level, 5. a large square; 6. a rod; 7. a jointing rule ; 8. a jointer; 9. a pair of compasses; 10. a raker; 11. a hod; 12. a pair of line pins; 13. a rammer ; 14. an iron crow ; 15. a pick axe ; 16. a grinding stone ; 17. a banker; 18. a camber slip ; 19. a rubbing stone ; 20. a bedding stone ; 21. a square ; 22. a bevel ; 23. a mould ; 24. a scribe ; 25. a saw; 26. an axe ; 27. a templet ; 28. a chopping block ; 29. a float stone.

---

§ 3. *A List of Tools used in Tiling.*

1. A lathing hammer ; 2. a laying trowel; 3. a boss ; 4. a pantile strike ; 5. a scurbage.

TOOLS FOR WALLING DESCRIBED.

§ 4. THE BRICK TROWEL

Is used for taking up mortar, and spreading it on the top of the walls, in order to cement together the bricks which are to be laid, and also to cut the bricks to any required lengths.

---

§ 5. THE HAMMER

Is used for cutting holes in brick-work.

---

§ 6. THE PLUMB RULE

Is about four feet long, with a line and plummet, in order to carry the faces of walls up vertically. See also Carpentry, § 14.

§ 7. THE LEVEL

Is about ten or twelve feet long, in order to try the level of walls at various stages of building, and particularly at window sills and wall plates. See also Carpentry, § 12, 13.

---

§ 8. THE LARGE SQUARE

Is used for setting out the sides of a building at right angles, which is also obtained by Prob. 1, 2, 3. Geometry, page 24.

---

§ 9. THE ROD

Is either five or ten feet in length, and used for measuring lengths, breadths, and heights, with more despatch than could be done by a pocket rule.

---

§ 10. THE JOINTING RULE

Is about eight or ten feet long, according to whether one or two bricklayers are to use it, and about four inches broad. By this rule, they run the joints of the brick-work.

---

§ 11. THE JOINTER

With which, and the jointing rule the horizontal and vertical joints are marked; it is shaped like the letter S, and is of iron.

---

§ 12. THE COMPASSES

Is used for traversing arches and vaults.

§ 13. THE RAKER

Is a piece of iron with two knees or angles, which divide it into three parts at right angles to each other; the two end parts are pointed and of equal lengths, and stand upon contrary sides of the middle part. Its use is to pick decayed mortar out of the joints in old walls, for the purpose of replacing the same with new mortar.

---

§ 14. THE HOD

Is a wooden trough, shut up at one end and open at the other, the sides consisting of two boards at right angles to each other; from the meeting of the two sides projects a handle at right angles: this machine is used by the labourer for carrying mortar and bricks; he strews the inner surface over with fine dry sand before he puts in the mortar, which prevents it sticking to the wood, then placing it upon his shoulder, carries the load to the bricklayer.

---

§ 15. THE LINE PINS

Are two iron pins for fastening and stretching the line, at proper intervals of the wall, in order to lay the course of brick-work level on the bed, and straight along the face of the wall. The line pins have generally a length of sixty feet of line, fastened to each pin.

---

§ 16. THE RAMMER

Is used for ascertaining whether the ground be sufficiently solid for building upon, also for beating the ground to a firm bearing, so as to give it the utmost degree of compression; for if ground is built upon in a loose state, in all probability fractures in the walls would ensue, and endanger the whole building. See Foundations.

No. 12.   y

§ 17. THE IRON CROW AND PICK AXE

Are used in conjunction for cutting or breaking through walls, or raising large and ponderous substances out of the ground, or the like.

___

§ 18. THE GRINDING STONE

Is used for sharpening axes, hammers, and other tools.

___

§ 19. THE BANKER

Is a bench from six to twelve feet in length, according to the number of those who are to work at it, and from two feet six inches, to three feet in breadth, and may be an inch thick, and raised about two feet eight inches from the ground. It is generally made of an old ledged door, set upon three or five posts in front, and its back edge against a wall. It is used for preparing the bricks for rubbed arches, or other gauged work upon.

___

§ 20. THE CAMBER SLIP

Is a piece of wood generally about half an inch thick, with at least one curved edge rising about one inch in six feet, for drawing the soffit lines of straight arches, when the other edge is curved, it rises only about one half of the other, viz. about half an inch in six feet, for the purpose of drawing the upper side of the said arch, so as to prevent it from becoming hollow by the settling of the arch. The upper edge of the arch is not always cambered, some persons preferring it to be straight. The bricklayer is always provided with a camber slip; which being sufficiently long, answers to many different widths of openings; when he has done drawing his arch, he gives the camber slip to the carpenter, in order to form the centre to the required curve of the soffit.

§ 21. THE RUBBING STONE

Is of a cylindric form, about twenty inches diameter, but may be more or less at pleasure, fixed at one end of the banker upon a bed of mortar. By this, the bricks which have been previously axed, are rubbed smooth; also the headers and stretchers in returns, which are not axed, called rubbed returns, and rubbed headers and stretchers.

§ 22. THE BEDDING STONE

Consists of a straight piece of marble, not less than eighteen or twenty inches in length, about eight or ten inches wide, and of any thickness. Its use is, to try the rubbed side of the brick, which you must first square, in order to prove whether the surface of the brick be straight, so as to fit it upon the leading skew back, or leading end of the arch.

§ 23. THE SQUARE

Is used in trying the bedding of the bricks, and squaring the soffits across the breadth of the said bricks.

§ 24. THE BEVEL

For drawing the soffit line on the face of the bricks.

§ 25. THE MOULD

Is used in forming the face and back of the brick, in order to its being reduced in thickness to its proper taper, one edge of the mould being brought close to the bed of the brick already squared; the mould has a notch for every course of the arch.

§ 26. THE SCRIBE

Is a spike or large nail ground to a sharp point, to mark the bricks on the face and back by the tapering edges of the mould, in order to cut them.

---

§ 27. THE TIN SAW

Is used for cutting the soffit lines about one-eighth part of an inch deep, first by the edge of the bevel on the face of the brick, then by the edge of the square on the bed of the brick, in order to enter the brick axe, and to keep the brick from spaltering. The saw is also used in cutting the soffit through its breadth, in the direction of the tapering lines, drawn upon the face and back edge of the brick, but the cutting is always made deeper on the face and back of the brick than in the middle of its thickness, for the said purpose of entering the axe : the saw is likewise used for cutting the false joints of headers and stretchers.

---

§ 28. THE BRICK AXE

Is used for axing off the soffits of bricks to the saw cuttings, and the sides to the lines drawn by the scribes. As the bricks are always rubbed smooth after axing, the more truly they are axed, the less labour there will be in rubbing.

---

§ 29. THE TEMPLET

Is used in taking the length of the stretcher and width of the header.

Note. The last ten articles relate entirely to the cutting of gauged arches, which are now the principal things that occur in gauged work.

§ 30. THE CHOPPING BLOCK

Is for reducing the bricks to their intended form by axing them, and is made of any chance piece of wood that can be obtained, from six to eight inches square, supported generally upon two fourteen-inch brick piers, provided only two men be to work at it but if four men, the chopping block must be lengthened and supported by three piers, and so on according to the number. It is about two feet three inches in height.

§ 31. THE FLOAT STONE

Is used for rubbing curved work smooth, such as the cylindrical backs and spherical heads of niches, so as to take out the axe marks entirely : but before its application, it must first be brought to the reverse form of the intended surface, so as to coincide with it, as nearly as possible, in finishing.

§ 32. *Of Cements.*

Calcarious cements may be classed according to the three following divisions: namely, simple calcarious cement, water cement, mastichs, or maltha.

1. Simple calcarious cement includes those kinds of mortar which are employed in land building, and consists of lime, sand, and fresh water.

Calcarious earths are converted into quick lime by burning, which being wetted with water, falls into an impalpable powder, with great extracation heat; and if in this state it is beat with sand and water, the mass will concrete and become a stony substance, which will be more or less perfect according to its treatment, or to the quality and quantities of ingredients. When carbonated lime has been thoroughly burnt, it is deprived of its water, and all or nearly all of its carbonic acid. Much of the water, during the process of calcination, being carried off in the form of steam.

Lime-stone loses about four-ninths of its weight by burning, and when fully burnt it falls freely, and will produce something more than double the quantity of powder or slacked lime in measure, that the burnt lime-stone consisted of.

Quick lime, by being exposed to the air, absorbs carbonic acid with less or greater rapidity, as its texture is more or less hard, and this by continued exposure, becomes unfit for the composition of mortar: and hence it is that quick lime made of chalk, cannot be kept for the same length of time between the burning and slacking, as that made from stone.

Marble, chalk, and lime-stone, with respect to their use in cements, may be divided into two kinds, simple lime-stone, or pure carbonate of lime, and argillo-ferruginous lime, which contains from one-twentieth to one-twelfth of clay and oxide of iron, previous to calcination: there are no external marks by which these can be distinguished from each other, but whatever may have been the colour in the crude state, the former, when calcined, becomes white, and the latter more or less of an ochery tinge. The white kinds are more abundant, and when made into mortar, will admit of a greater portion of sand than the brown; consequently, are more generally employed in the composition of mortar; but the brown lime is by far the best for all kinds of cement. If white, brown, and shell lime recently slacked, be separately beat up with a little water into a stiff paste; it will be found that the white lime, whether made from chalk, lime-stone, or marble, will not acquire any degree of hardness; the brown lime will become considerably indurated, and the shell lime will be concreted into a firm cement, which, though it will fall to pieces in water, is well qualified for interior finishings, where it can be kept dry.

It was the opinion of the ancients, and is still received among our modern builders, that the hardest lime-stone furnishes the best lime for mortar; but the experiments of Dr. Higgins and Mr. Smeaton, have proved this to be a mistake; and that the softest chalk lime, if thoroughly burnt, is equally durable with the hardest stone-lime, or even marble : but though stone and chalk lime are equally good, under this condition, there is a very important practical difference between them, as the chalk lime absorbs carbonic acid with much greater avidity; and if it is only partially calcined, on the application of water it will fall into a coarse powder, which stone lime will not do.

For making mortar, the lime should be immediately used from the kiln, and in slacking it, no more water should be allowed than what is just sufficient: and for this purpose, Dr. Higgins recommends lime water.

The sand made use of should be perfectly clean ; if there is any mixture of clay or mud, it should be divested, of either or both, by washing it in running water. Mr. Smeaton has fully shown by experiment, that mortar, though of the best quality, when mixed with a small proportion of unburnt clay, never acquires that hardness, which without this addition, it speedily would have attained. If sea sand is used, it requires to be well washed with fresh water, to dissolve the salt with which it is mixed, otherwise the cement into which it enters, never becomes thoroughly dry and hard: the sharper and coarser the sand is, the stronger is the mortar, also a less proportion of lime is necessary. It is therefore more profitable to use the largest proportion of sand, as this ingredient is the cheapest in the composition.

The best proportion of lime and sand in the composition of mortar, is yet a desideratum.

It may be affirmed in general, that no more lime is required to a given quantity of sand, than what is just sufficient to surround the particles, or to use the least lime so as to preserve the necessary degree of plasticity. Mortar in which sand predominates, requires less water in preparing, and therefore sets sooner; it is harder and less liable to crack in drying, for this reason, that lime shrinks greatly in drying, while sand retains its original magnitude. We are informed by Vitruvius, *lib.* 2. *c.* 5. that the Roman builders allowed three parts of pit sand, or two of river or sea sand, to one

of lime; but by Pliny, (Hist. Nat. *lib.* xxxvi.) four parts of coarse
sharp pit sand, and only one of lime. The general proportion
given by our London builders, is one hundred weight and a half, or
thirty-seven bushels of lime and two loads and a half of sand; but
if proper caution were taken in the burning the lime, the quality
of the sand, and in tempering the materials, a much greater quantity of sand might be admitted.

Mr. Smeaton observes, that there is scarcely any mortar, that
if the lime be well burnt, and the composition well beat in the
making, but what will require two measures of sand, to one or
unslacked lime; and it is singular, that the more the mortar is
wrought or beat, a greater proportion of sand may be admitted.
He found that by good beating, the same quantity of lime would
take in one measure of tarras, and three of clean sand, which
seems to be the greatest useful proportion.

Dr. Higgins found that a certain proportion of coarse and fine
sand, improved the composition of mortar; the best proportion of
ingredients, according to experiment made by him, are as follow,
by measure :

Lime newly slacked • • •	1 part.
Fine sand - • • • •	3 parts
Coarse sand - • • • •	4 parts.

He also found that an addition of one-fourth part of the quantity
of lime, of burnt bone ashes, improved the mortar by giving the
tenacity, and rendering it less liable to crack in drying.

The mortar should be made under ground, and then covered up
and kept there for a considerable length of time, the longer the
better; and when it is used, it should be beat up afresh. This
makes it set sooner, renders it less liable to crack, and more
hard when dry.

The stony consistence which it acquires in drying, is owing to
the absorption of carbonic acid, and a combination of part of the
water with the lime : and hence it is that lime that has been long
kept after burning, is unfit for the purpose of mortar; for in the
course of keeping, so much carbonic acid has been imbibed, as to

have little better effect in a composition of sand and water, than chalk or lime-stone reduced to a powder from the crude state would have in place of it.

Grou is mortar containing a larger proportion of water than is employed in common mortar, so as to make it sufficiently fluid to penetrate the narrow irregular interstices of rough stone walls. Grout should be made of mortar that has been long kept and thoroughly beat; as it will then concrete in the space of a day: whereas if this precaution is neglected, it will be a long time before it set, and may even never set.

Mortar made of pure lime sand and water, may be employed in the linings of reservoirs, and aqueducts, provided that it has sufficient time to dry; but if the water be put in while it is wet, it will fall to pieces in a short time, and consequently, if the circumstances of the building are such as render it impracticable to keep out the water, it should not be used; there are, however, certain ingredients put into common mortar, by which it is made to set immediately under water, or if the quick lime contain in itself a certain portion of burnt clay, it will possess this property.

This is all that is necessary to say under this head; what relates to mortars employed in aquatic buildings will be treated of under water cements.

From the friable and crumbling nature of our mortar, a notion has been entertained by many persons, that the ancients possessed a process in making their mortar, which has been lost at the present day; but the experiments of Mr. Smeaton, Dr. Higgins, and others, have shown this notion to be unfounded; and that nothing more is wanting, than that the chalk, lime-stone, or marble, be well burnt and thoroughly slacked immediately, and to mix it up with a certain proportion of clean, large-grained, sharp sand, and as small a quantity of water as will be sufficient for working it; to keep it a considerable time from the external air, and to beat it over again before it is used: the cement thus made will be sufficiently hard.

The practice of our modern builders is to spare their labour,
z

and o increase the quantity of materials they produce, without
any regard to its goodness; the badness of our modern mortar
is to be attributed both to the faulty nature of the materials, and
to the slovenly and hasty methods of using it. This is remarkably
instanced in London, where the lime employed is chalk lime, in-
differently burnt, conveyed from Essex or Kent, a distance of ten
or twenty miles, then kept many days without any precaution to
prevent the access of external air: now in the course of this time,
it has absorbed so much carbonic acid as nearly to lose its cementing properties; and though chalk lime is equally good with the
hardest lime-stone, when thoroughly burnt, yet by this treatment,
when it is slacked, it falls into a thin powder, and the core or
unburnt lumps are ground down, and mixed up in the mortar, and
not rejected as they ought to be.

The sand is equally defective, consisting of small globular
grains, containing a large proportion of clay, which prevents it
from drying, and attaining the necessary degree of hardness.

These materials being compounded in the most hasty manner,
and beat up with water in this imperfect state, cannot fail of producing a crumbling and bad mortar. To complete the hasty hash,
screened rubbish, and the scraping of roads, also are used as sub-
stitutes for pure sand.

How very different was the practice of the Romans! the lime
which they employed, was perfectly burnt, the sand sharp, cleaned,
and large grained: these ingredients were mixed in due proportion
with a small quantity of water, the mass was put into a wooden
mortar, and beat with a heavy wooden or iron pestle, till the composition adhered to the mortar; being thus far prepared, they kept
it till it was at least three years old. The beating of mortar is of
the utmost consequence to its durability, and it would appear that
the effect produced by it, is owing to something more than a mere
mechanical mixture.

Water cements are those which are impervious to water, generally made of common water, or of pure lime and water, with the

addition of some other ingredient which gives it the property of hardening under water.

For this purpose, there are several kinds of ingredients that may be used.

That known by the name of pozzolana, which is supposed to consist of volcanic ashes thrown out of Vesuvius, has been long celebrated, from the early ages of the Romans, to the present day. It seems to consist of a ferruginous clay, baked and calcined by the force of volcanic fire; it is a light, porus, friable mineral, of a red colour. The cement employed by Mr. Smeaton, in the construction of the Eddystone light-house, was composed of equal parts by measure, of slacked aberthaw lime and pozzolana; this proportion was thought advisable, as this building was exposed to the utmost violence of the sea; but for other aouatic works as locks, basins, canals, &c. a composition made of lime, pozzolana, sand, and water, in the following proportion: viz. two bushels of slacked aberthaw lime, one bushel of pozzolana, and three of clean sand, has been found very effectual.

---

§ 33. *Description of Bricks.*

Bricks are a kind of factitious stone, composed of argillaceous earth, and frequently a certain portion of sand, and cinders of sea-coal, tempered together with water, dried in the sun, and burnt in a kiln, or in a heap or stack called a clamp.

Bricks are first formed from the clay into rectangular prisms, in a mould of ten inches in length, and five in breadth; and when burnt, usually measure nine inches long, four and a half broad, and two and a half thick: so that a brick generally shrinks one inch in ten; but the degree of shrinking is not always the same, it depends upon the purity and tempering of the clay, and also upon the burning.

For brick-making, the earth should be of the purest kind, dug in autumn, and exposed during the winter's frost; this allows the air to penetrate, and divide the earthly particles, and facilitates the subsequent operations of mixing and tempering.

To make real good bricks, the earth should be dug two or three years before it is used, in order to pulverize it; and should be mixed with a due proportion of clay and sand, as too much argillaceous matter causes the bricks to shrink, and too much sand renders them heavy and brittle. The London practice of mixing sea-coal ashes, and in the country light sandy earth, not only makes them work easy and with greater despatch, but tends also to save coals or wood in burning them. The earth should be entirely divested of stony particles, and should be often beat or turned over, with as little water as possible, in order to incorporate the soil with the ashes or sand, until the whole be converted into an uniform paste; and note, that too much water prevents the adhering of the parts; before the bricks are burnt, they should be thoroughly dry, or they will crack and crumble in the burning.

Bricks made of good earth, well tempered, become solid, smooth, hard, durable, and ponderous; but require half as much more earth, also a longer time in drying and burning them, than common bricks, which are light, spongy, and full of cracks. Bricks are either burnt in clamps or kilns; the former is the practice about London, and the latter in the country; bricks burnt in kilns, are less liable to waste, require less fuel, and are sooner burnt than in clamps. It must be observed, that steeping of bricks in water, after once burning, and then burning them afresh, makes them more than doubly strong.

There are several kinds of bricks, as marls, stocks, and place brick. The only difference in making them is, that marls are prepared and tempered with the greater care; the construction of the clamp is the same for each, but for marls, greater care is taken not to over-heat the kiln, but that it burn equally and moderately, and as diffusively as possible. The finest kind of marls called firsts, are selected, and used as cutting bricks, for arches, over doors, windows, and quoins, for which they are gauged are rubbed to their proper forms. The next best called seconds, are selected and used for principal fronts.

Marls are every way superior to stock bricks, not only in colour, which is a pleasant pale yellow, but also in point of smoothness and durability. Hence the gray stocks are an inferior kind. The place bricks, or as they are otherwise called peckings, and sometimes sandal or semel bricks, are those that are left of the clamp after taking away the rubbers and marls; their inferior quality is occasioned by not being sufficiently and uniformly burnt: they also differ from stock bricks in being of a redder colour, and of a more uneven texture. Burrs are over-burnt brick, sometimes two or three are quite vitrified and run together. There are also red stocks; these are made in the country, and burnt in kilns; the best kind are used as cutting bricks, and are called red rubbers. Fine bricks are made at Hedgerly, a village near Windsor, and are therefore also called Windsor bricks. These are very hard, of a red colour, and will stand the utmost fury of the fire; their length and breadth are the same as stock bricks, but their thickness is only about one inch and a half. Bricks are sold by the thousand. Stock and place bricks made for sale, shall not be less than eight inches and a half long, four inches wide, and two inches and a half thick, when burnt, by 17 Geo. iii. *cap.* 69.

Besides the bricks of our own manufacture, Dutch clinkers are also imported for the purpose of paving yards and stables. These are very hard, of a brimstone colour, and almost vitrified in burning. They are about six inches long, three broad, and one thick, and look extremely well when laid herring-bone ways.

As a building material, bricks have several advantages over stone, being lighter, and from their porus structure they unite better with the mortar, and are not so liable to attract damp.

Bricks for paving floors, also called paving tiles, are of several magnitudes, and are made of a stronger clay. The largest are about twelve inches square, and one and a half in thickness; the second are about nine inches square, though called ten, being

formerly so, and one and a quarter thick; these may be rubbed
smooth, and when laid diagonally, have a very pleasing effect.
Bricks for paving are about nine inches long, four and a half
broad, and one and a half thick.

The chief covering for roofs in and about London is slate:
however, in the interior of the country, tiles are almost uniformly
used for the roofs of houses, and in some instances on barns; tiles
for roofs are of several kinds, as pan tiles, plain tiles, ridge tiles,
and hip tiles.   Pan tiles are about 13 inches long, 8 inches broad,
and about half an inch thick; their transverse section is a figure of
contrary curvature, the form of the tile being two portions of cylindric surfaces on both sides; the part which is of the greatest
radius serves as a channel for discharging the rain water, and the
other part, which is of much less radius, serves to lap over the
edge of the adjoining tile : at the upper end of the tile projects a
knob from the under and convex side, for the purpose of hanging
it to the laths.   The laths used for pan tiles are about three quarters
of an inch thick, and one and a quarter broad, made of deal.   Flemish tiles are sometimes imported from Holland; they are very
hard and durable, and are glazed of a leaden colour.

---

§ 34. *Foundations.*

Having dug the trenches for the foundations, the ground must
be tried with an iron crow, or with a rammer, and if found to shake
it must be pierced with a borer, such as is used by well diggers :
then if the ground proves to be generally firm, the loose or soft parts
if not very deep, must be excavated until a solid bed appears; but
observe in building up these parts that the bottom of the excavation
must widen upwards in a gradual slope, in the direction of the
trenchers, in form of a series of steps, which will admit of a firmer
bed for the stones, so that they will have no tendency to slide, as
would be the case if built upon inclined planes : and thus in wet
seasons, the moisture in the foundations would induce the inclined

parts to slide, and descend by their gravity towards the lowest parts, and in all probability would fracture the walls, and endanger the whole fabric.

If the ground proves soft in several places to a great depth under apertures, and firm upon the sides on which the piers between the windows of the superstructure are to be erected, the better way is to turn inverted arches under the apertures, (see Plate XXVI.) and indeed at all times where there is sufficient height of wall below the apertures to admit of them, it is a necessary precaution.

For the small base of the piers will more easily penetrate the ground than one continued base: and as the piers are permitted to descend in a certain degree, and so long as they can be kept from spreading, will carry the arch along with them, and compress the ground, which will therefore re-act against the under sides of the inverted arch, which, if closely jointed will not yield, but act with the abutting piers as one solid body. On the contrary, if no inverted arches were used, the low piece of walling under the apertures not having a sufficient vertical dimension would give way from the resistance of the ground upon its base, and thereby, not only fracture the spaces of brick-work which lie vertical between the apertures, but breaks the sills of the windows. Where the precaution of inverted arches is omitted, and the building is weighty, the probability of the event of fracturing the walls is almost certain; the author, who has had great practice in conducting buildings, never experienced any instance to the contrary, in the numerous buildings in which he has been concerned. It is therefore of the utmost consequence to throw these arches with the greatest care; they ought not to be less in height than half their width, and as a parabolic curve is very easily described, it would be still more effectual in resisting the re-action of the ground than one of uniform curvature, as the arc of a circle; the parabolic arch or vault being the form adapted more nearly to the laws of uniform pressure. From the equality of the curvature of the circle, it is only capable of resisting a uniform pressure upon all points directed to the centre, and thus a cylindric vessel surrounded with water is a proper form of a hollow body to be con-structed of the least quantity of materials, or at the least expense.

The bed of the piers ought to be as uniform as possible, for though all the parts of the bottom of the trenches may be very firm, if there be any difference, as they will all sink, the quantity which they will give will be according to the softness of the ground, therefore the piers erected upon the softer, will descend more than those on the firmer ground, and occasion a vertical fracture in the building.

If the hard parts of the foundation are only to be found under apertures, then build piers in these places, and instead of inverted arches suspend arches between the piers. In the construction of the arches some attention must be paid to the breadth of the in-sisting pier, whether it will cover the arch or not : for suppose the middle of the piers to rest over the middle of the summit of the arches, then the narrower the piers, the more curvature the sup-porting arch ought to have at the apex. When arches of suspen-sion are used, the intrados ought to be clear, so that the arch may have the full effect; but as observed before, it will also be requi-site here, that the ground be uniformly hard on which the piers are erected, for the reasons already given ; but it might be farther observed, that even where the ground is not very hard under the piers, if it is but uniform, the parts will descend equally, and the building will remain uninjured.

If the foundation be not very insufficient, it may be made good by ramming large stones closely laid with a heavy rammer, of a breadth at the bottom proportioned to the insisting weight, and this breadth in ordinary cases may project a foot on each side of the wall, then another course may be laid upon this so as to bring the upper bed of the stones upon a general level with the trenches, and to project about eight inches on each side of the wall, or to recede four inches on each side within the lower course. In lay-ing of these courses, care should be taken to chop or hammer-dress the stones, so as to have as little taper as possible, and to make the joints of the one course fall as nearly to the middle of the stones in the adjoining course as possible; and this principle must be strictly adhered to in all walling whatever; and though there are various modes of disposing stones or bricks, the end is to obtain the greatest uniform lap upon each other, throughout the whole.

If the foundation is very bad, the whole must be piled as already described in the department of Carpentry.

---

§ 35. *Walls.*

We shall now suppose that the ground is either naturally sufficient for building upon, or is prepared for the purpose by means similar to what have already been described : and the different qualities of mortar and bricks being also described: such materials must be employed in the construction of the whole, or in the different parts, as are sufficient for the end proposed; thus, in places exposed to the weather, more durable materials must be employed than in those which are covered; but in this, some regard must also be had to the importance of the fabric, or whether long duration may be required or not.

When you slack the lime, wet it only with so much water as is sufficient to reduce it to a powder, and only about a bushel at a time, covering it over with a layer of sand, in order to prevent the gas, which is the virtue of the lime, from escaping. The best proportion of the ingredients of lime and sand for mortar has been fully specified, but in ordinary cases, where time will not permit to prepare the materials to the best advantage, or where the end proposed would not be a compensation for the expense, about two or two and a half measures of sand to one of lime may be used; but even this proportion will not always hold, for some lime will require more and some less sand; this being understood, slack the same quantity of lime alternately, until the whole is made up: this is a better mode than to slack the whole at once, as the exposure is less in the former, than in the latter case.

Nos. 13 & 14.   2 A

Beat your mortar with the beater three or four times over before it is used, so as to incorporate the lime and sand, and to break the knots that pass through the sieve; this will not only render the texture uniform, but will make the mortar much stronger by permitting the air to enter the pores: and observe here also, as we have before stated, to use as little water in the beating as possible. Should the mortar stand any time after beating, it should be beat again immediately before it is used, so as to give tenacity and to prevent labour to the bricklayer. In summer dry hot weather use your mortar pretty soft, but in winter rather stiff.

If you lay your bricks in dry weather, and if you require firm work, you must use mortar prepared in the best way, and before using the bricks, they must be wetted or dipped in water as they are laid on the wall, but in moist weather this will be unnecessary. The wetting of the bricks causes them to adhere to the mortar, whereas, if laid dry, and covered with sand or dust, they will never stick, but may be taken off without the adhesion of a single particle of mortar.

In winter, as soon as the frost or stormy season begins to set in, the walls must be covered; for this purpose straw is usually employed, and sometimes in particular buildings a capping of weather boarding, in form of a stone coping, for throwing the water equally to both sides, is used; but even in this case, it would be better to have straw under the wood, which would be still a farther proof against frost. There is nothing so prejudicial to a building as alternate rain and frost, if exposed; for the rain makes way through the pores into the heart of the stone and mortar, and when the freezing comes on, the water is converted into ice, which expands beyond the original bulk with such power, that no known force of compression is capable of preventing it from expansion. In consequence of this, the heaviest stones and even the largest rocks have been burst. Though this is the cause why buildings decay in lapse of time, yet the vertical surfaces exposed to the weather suffer but in an incomparably small degree to horizontal surfaces thus exposed.

In working up the wall it would be proper not to work more than four or five feet at a time, for as all walls immediately after building shrink, the part which is first brought up will remain stationary, and when the adjacent part is also brought up, it will shrink in altitude by itself, and consequently will separate from the other which has already become fixed. In carrying up any particular part, the ends should be regularly sloped off so as to receive the bond of the adjoining parts on the right and left. There is nothing that will justify one part of a wall being carried higher than one scaffold, except it be to forward the carpenter in some particular part, or the like.

In brick work there are two kinds of bond, one in which a row of bricks laid lengthways in the length of the wall, is crossed by another row laid with their breadth in the said length, and thus proceeding to work up the courses in alternate rows, which is called English bond. The courses in which the length of the bricks are disposed in the length of the wall are called stretching courses, and the bricks themselves are called stretchers. The courses in which the length of the bricks run in the thickness of the wall are called heading courses, and the bricks thus disposed are called headers. The other kind of brick work is the placing of header and stretcher alternately in the same course; this disposition of the bricks is called Flemish bond. This latter mode, though esteemed the most beautiful, is attended with great inconveniences in the execution, and in most cases is incapable of uniting the parts of a wall with the same degree of firmness as the English bond

To enter into the particular merits of these two species of bond would carry this department beyond its allowed limits; the reader who wishes farther satisfaction will consult the explanation of the Plates, and an ingenious tract on *Brick Bond*, by Mr. G. Saunders, where the defects of Flemish bond, and the superiority of the old English bond, are pointed out in the most satisfactory manner. However, it may be proper to observe in general, that whatever advantages are gained by any disposition of placing the bricks in Flemish bond in the particular direction, is lost in another: thus, if an advantage is gained in tying a wall together in its thickness, it is lost in the longitudinal bond, and the contrary. In order to remedy this inconvenience in thick walls, some place the bricks in the core at an angle of forty-five degrees, and parallel to each other throughout the length of each course, so as to cross each other at right angles in the succeeding course: but even the advantages obtained by this disposition are not satisfactory, for though those bricks in the middle of the core have sufficient bond, yet where they join to the bricks on the sides of the wall, they form triangular interstices, and therefore the sides must be very imperfectly tied to the core.

§ 36. *Vaulting and Groining.*

DEFINITIONS.

A simple vault is an interior concavity extended over two parallel opposite walls, or over all diametrically opposite sides of one circular wall.

The concavity or interior surface of the vault is called the intrados.

The intrados of a simple vault is generally formed of the portion of the surface of a cylinder, cylindroid, or sphere, never greater than that of half the solid, and the springing lines which terminate the walls that the vault rises from, are generally straight lines, parallel to the axis of the cylinder or cylindroid.

When the vault is spherical, the circular wall terminates in a level plane at top from which the vault springs, and forms either a complete hemisphere, or a portion of the sphere less than the hemisphere.

Conic surfaces are seldom employed in vaulting, but when a conic surface is employed for the intrados of a vault, it should be semi-conic with a horizontal axis, or the surface of the whole cone with its axis vertical.

All vaults which have a horizontal straight axis, are called straight vaults.

All vaults which have their axis horizontal, are called horizontal vaults.

A groin is the excavation or hollow formed by one simple vault piercing another, or a groin is that in which two geometrical solids may be transversely applied one after the other, so that a portion of the groin will have been in contact with the first solid, and the remaining part in contact with the second solid, when the first is removed. The most usual kind of groining is one cylinder piercing another, or a cylinder and cylindroid piercing each other, having their axis at right angles.

The axis of each simple vault forming the intrados of a groin is the same with the axis of the geometrical solids, of which the intrados of the groin is composed.

When the breadths of the cross pages or openings of a groined vault are equal, the groin is said to be equilateral.

When the altitudes of the cross vaults are equal, the groin is said to be equi-altitudinal.

Groins have various names, according to the surfaces of the geometrical bodies, which form the simple vault.

A cylindric groin is that which is formed by the intersection of one portion of a cylinder with another.

A cylindroidic groin is that which is formed by the intersection of one portion of a sphere with another.

A spheric groin is that which is formed by the intersection of one portion of the sphere with another.

A conic groin is that which is formed by the intersection of one portion of a cone with another.

The species of every groin formed by the intersection of two vaults of unequal heights, is denoted by two preceding words, the former of which ending in o, indicates the simple vault which has the greater height, and the latter ending in i c indicates the simple vault of the less height.

When a groin is formed by the intersection of two unequal cy
p 2

lindric vaults, it is called a cylindro-cylindric groin, and each arch so formed is called a cylindro-cylindric arch.

When a groin is formed by the intersection of a cylindric vault with a spheric vault, and the spheric portion being of greater height than the cylindric portion, the groin is called a sphero-cylindric groin, and each arch forming the groin is called a sphero-cylindric arch.

When a groin is formed by the intersection of a cylindric vault, with a spheric vault, and the spheric portion of less altitude than the cylindric portion, it is called a cylindro-spheric groin; and each arch forming the groin is called a cylindro-spheric arch.

When one conic vault pierces another of greater altitude, the groin formed by the intersection is called a cono-conic groin; and each arch forming the groin, a cono-conic arch.

A rectangular groin is that which has the axis of the simple vault in two vertical planes, at right angles to each other.

A multangular groin is that which is formed by three or more simple vaults piercing each other, so that if the several solids which form each simple vault be respectively applied, only one at a time to succeeding portions of the groined surface, every portion of the groined surface will have formed successive contact with certain corresponding portions of each of the solids.

An equi-angular groin is that in which the several axes of the simple vaults form equal angles, around the same point, in the same horizontal plane.

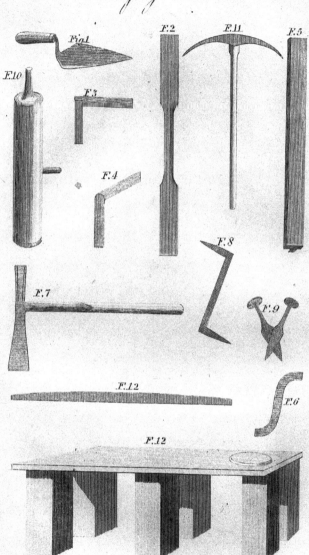

Bricklaying Plate XXIII.

Fig.1

F.10 F.3 F.2 F.11 F.5

F.4

F.8

F.7 F.9

F.12

F.6

F.12

W.S.Barnard, Sculp.

Bricklaying Plate 24

Fig 1

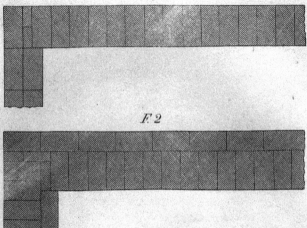

F. 2

F. 3

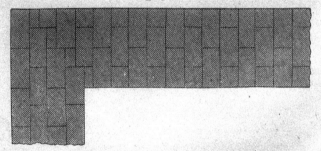

F. 4

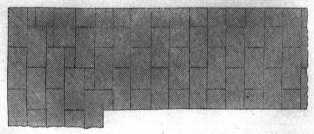

§37. EXPLANATIONS OF THE PLATES IN
BRICKLAYING.

PLATE XXIII.

Fig. 1 the brick trowel.
Fig. 2 the brick axe.
Fig. 3 the square.
Fig. 4 the bevel.
Fig. 5 the jointing rule.
Fig. 6 the jointer.
Fig. 7 the hammer.
Fig. 8 the raker.
Fig. 9 the line pins.
Fig. 10 the rammer.
Fig. 11 the pick axe.
Fig. 12 the camber slip.
Fig. 13 the banker, with the rubbing stone placed at one end

PLATE XXIV.

Various specimens of English bond according to the different
thicknesses of walls ; in these the heading and stretching courses
mutually cross each other in the core of the wall, and therefore
produce an equality of strength.

Fig. 1 shows the bond of a nine inch wall: here as well as in
the following it must be observed, that as the longitudinal extent
of a brick is nine inches, and the breadth four and a half inches,
in order to prevent two vertical joints from running over each
other at the end of the first stretcher from the corner, after placing
the return corner stretcher, which becomes a header in the face
that the stretcher is in below, and occupies half the length of
this stretcher ; a quarter brick is placed upon the side, so that the

two together extend six inches and three quarters, and leave a lap
of two inches and a half for the next header, which being laid,
lies with its middle upon the middle of the header below, and in
this manner the bond is continued. The brick-bat thus introduced
next to the corner header is called a closer. The same effect
might be obtained by introducing a three-quarter bat at the corner
in the stretching course, for then when the corner header comes
to be laid over it, a lap of two inches and a quarter will be left at
the end of the stretchers below for the next header, which being
laid, the joint below the stretchers will coincide with its middle,
and in this manner the bond may be continued as before.

Fig. 2 a fourteen inch, or brick and half wall. In this the
stretching course upon the one side is so laid, that the middle of
the breadth of the bricks in the heading course upon the opposite
side falls alternately upon the middle of the stretchers, and upon
the joints between the stretchers.

Fig. 3 a two brick wall. In the heading course, every alternate header is only half a brick thick on both sides in order to
break the joints in the core of the wall.

Fig. 4 a two brick and a half wall, bricks laid as in Fig. 3.

––––––

PLATE XXV.

Contains various specimens of Flemish bond according to the
different thicknesses of walls. The dotted lines show the disposition of the bricks in the courses above.

Fig. 1 a nine inch wall where two stretchers lie between two
headers, the length of the headers and the breadth of the stretchers
extending the whole thickness of the wall.

Fig. 2 a brick and a half wall, one side being laid as in Fig. 1,
and the opposite side, with a half header opposite to the middle of
the stretcher, and the middle of the stretcher opposite the middle
of the end of the header.

Fig. 3 another disposition of Flemish bond where the bricks

Fig 1

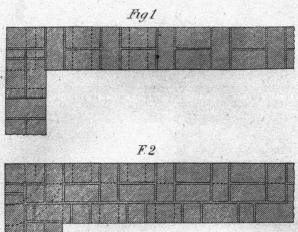

F. 2

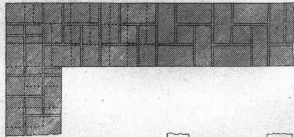

F. 3

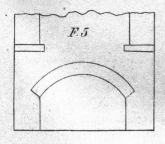

F. 4

F. 5

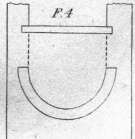

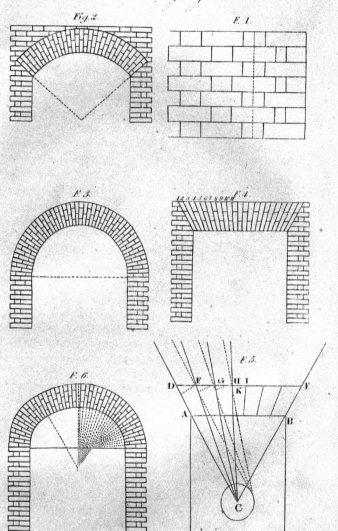

Fig. 2.

F. 1.

F. 3.

F. 4.

F. 6.

F. 5.

are similarly disposed on both sides ot the wall, the tail of the headers being placed contiguous to each other, so as to form square spaces in the corner of the wall for half bricks.

Fig. 4 a reversed arch supposed to come under a window, in order to prevent the fracturing of the wall under the lowest window. Arching under the apertures should never be omitted in any building whatever, provided there be room ; if not, pieces of timbers ought to be laid, so as to present the most inflexibility to the ground, and make the wall act longitudinally as one solid body.

Fig. 5 supposed to be the case where the ground stands firm under the apertures, the weight of the pier is therefore discharged from the soft part under the piers. In this case if the bond of the pier is good, there will be very little danger of the wall fracturing under the apertures.

PLATE XXVI.

Fig. 1 part of the upright of a wall, at the return, laid with Flemish bond.

Fig. 2 a scheme arch, being two bricks high.

Fig. 3 a semi-circular arch two bricks high.

Fig. 4 a straight arch, which is usually the height of four courses of brick work : the manner of describing it will be shown in the following figure.

Fig. 5. To draw the joints of a straight arch, let A B be the width of the aperture; describe an equilateral triangle A B C upon this width ; describe a circle around the point C equal to the thickness of the brick. Draw D E parallel to A B at a distance equal to the height of four courses, and produce C A and C B to D and E. Lay the straight edge of a rule from C to D, and with a pair of compasses, opened to a distance equal to the thickness of a brick, cross the line D E at F, removing the rule from the points

2 B

C and D. Place the straight edge against the points C and F, and with the same extent, between the points of the compass, cross the line D E at G : proceed in this manner until you come to the middle, and as it is usual to have a brick in the centre to key the arch in, if the last distance which we will suppose to be H I is not equally divided by the middle point K of D E, the process must be repeated till it is found to be so.

Though the middle brick tapers more in the same length than the extreme bricks, it is convenient to draw all the bricks with the same mould, which is a great saving of time, and though this is not correctly true, the difference is so trifling as not to affect the practice. It may however be proper to observe, that the real taper of the mould is less than in the middle, but greater than either extreme distance : but even the difference between this is so small, that either may be used, or taking half their difference will come very near the truth. This difference might easily be shown by a trigonometrical calculation, the middle being an isosceles triangle, of which the base and perpendicular are given, the base being a certain part of the top line. In the triangle upon the sides you have one angle equal to 60 degrees, and the side D F is given, and $D C = (D K^2 \times K C^2)$ one half, can easily be found, so that in this triangle the two sides and the contained angle are given.

Fig. 6 an elliptic arch, the top is divided into equal parts, and not the underside.

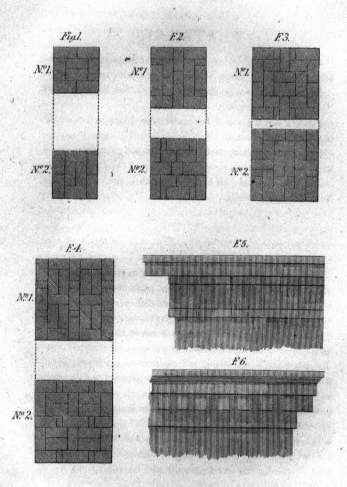

Bricklaying. Plate XXVII.

Fig.1.   F.2.   F.3.

N.º1.   N.º1.   N.º1.

N.º2.   N.º2.   N.º2.

F.4.   F.5.

N.º1.

N.º2.   F.6.

PLATE XXVII.

Contains piers of various substances according to the Flemish bond disposition of bricks, with designs of brick cornices.

Fig. 1 a pier, two brick square: No. 1. the bottom course, No. 2. the upper course

Fig. 2 a two and a half brick pier: No. 1. the bottom course, No. 2. the upper course.

Fig. 3 a three brick pier: No. 1. the bottom course, No. 2. the upper course.

Fig. 4 a three and half brick pier : No. 1. the bottom course, No. 2. the upper course.

Ornamental Brick Cornices.

In the construction of any thing destined to answer a particular end, it frequently happens that different kinds of materials may be employed for the purpose : it is evident that every distinct species of material will require its own peculiar manner of treatment, and the sizes of the parts which are to compose the thing required, must depend upon what the material will most conveniently admit of: thus brick, wood, stone, or iron, may be employed to con- struct a body for any proposed end : the manner of working these will not only differ, but the sizes of the things which are to com- pose the whole, and not only so, but sometimes a change in the general form also.

In brick cornices, from the various kinds of bricks and tiles, a variety of pleasing symmetry may be formed by various disposi- tions of the bricks, and frequently without cutting, or if cut, chamfering only may be used.

Fig. 5 a cornice in imitation of the Grecian Doric.

Fig. 6 a dentil cornice , in this last the upper member is cham- fered to give it the appearance of a moulding.

PLATE XXVIII.

Contains groins of various kinds.

Fig. 1 a semi-cylindric equi-angular groin, the centre of one vault being generally boarded in without any regard to the other, and the other boarded in afterwards.

Fig. 2 a cylindroidic-cylindric groin, being the intersection of a cylinder with a cylindroid.

Fig. 3 a cylindro-cylindric groin, being the intersection of one cylinder with another, and the cylindro vault being the highest.

Fig. 4 an improvement to the common four-sided groin, by Mr. Tappen, Architect, by raising the angles from an octagonal pier, instead of a square one; by this means, the pier may be made equally strong, by giving it more substance, and cutting away the angles will be more commodious for the turning any kind of goods round the corner; this may therefore be looked upon as a very considerable improvement in the vaultings of cellars of warehouses. This convenience is not the only improvement which this construction admits of, but the angles of the groin are strengthened by carrying the band round the diagonals of equal breadth, which affords better bond to the bricks, which are usually so much cut away, that instead of giving support, are themselves supported by the adjacent filling-in arches.

Fig. 5 the centreing for an hexagonal Gothic groin, such as are frequently seen in chapter houses.

Fig. 6 the piers of an hexagonal groin, and the angles obtunded according to the plan of Mr. Tappen. This construction is purely Gothic, the springers would cover the obtunded parts of the groined angles, and columnar mouldings those of the piers.

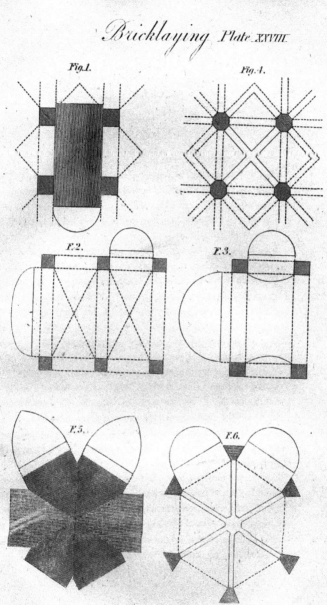

Bricklaying Plate XXVIII

Fig.1.

Fig.4.

F.2.

F.3.

F.5.

F.6.

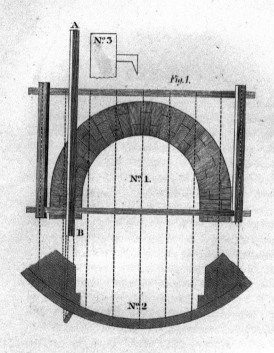

Fig. 1.

A

Nº 3

Nº 1.

B

Nº 2

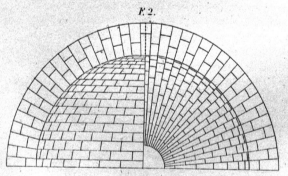

F. 2.

W.S. Barnard Sc.

PLATE XXIX.

The method of cutting the bricks for a cylindro-cylindric arch, and two different methods for the joints of the heads of niches.

Fig. 1 the cylindro-cylindric arch, with a frame of wood so constructed, that the two horizontal pieces have their outer edges in circles concentric with the circle of the wall: this is shown by the plan of the wall No. 2. The edges of the circular pieces are graduated with divisions perpendicularly over each other, A B: No. 2. is a rule to be moved vertically along the said concentric edges, which vertical position is always known by the corresponding divisions; on the front edge of the rule is a hook projecting so as to come to the cylindric surface of the wall: the hook is shown at No. 3, with a part of the rule. The use of this machine is for drawing the edges of the bricks in order to cut them to the circle.

Fig. 2 two different methods of forming the joints for the heads of spherical niches. In the right hand half the joints run horizontally, but this is a very bad method; as all the beds are conical, the bricks at the summit have little or no hold. In the other half the joints run radially in planes from the face to the centre. The work is not only more firm when executed by this last method, as bedding the courses on planes, but much more easily executed; nothing is more difficult to form than conical surfaces; and in this both conical and spherical surfaces occur; whereas when the joints run radially, only the spheric surface occurs, which may be formed by one bevel, only one side being straight and the other circular.

PLATE XXX.

Shows the method of steening wells.

The first thing is to make a centre, which consists of a boarding of inch or inch and a half stuff, ledged within with three circular rings. The bricks are laid between these rings, and all headers. The wide joints next to the boarding are filled in with tile or broken bricks. Where the soil is firm, centreings are not necessary, but they are requisite in sandy ground. The centreing remains permanently with the brick work; as the well digger excavates the soil, the first centre sinks, then a second centre is made, and put above the first, and built in with brick work in the same manner: and thus the number of centreings depend on the depth of the well. This method is that used in London: but in the country other methods are used. One is with several rings of timber without the boarding: they first build upon the first ring, four or five feet, then a second ring, and build again, and so on to the depth of the well. This however is not so good a method as the foregoing, as the sides of the brick work are very apt to bulge, particularly if great care be not taken in filling and ramming the sides in uniformly, so as to press equally at the same time.

———

Abstract of the Building Act, as far as regards the Bricklayer, 14 Geo. III. *which refers only to London, and the several Parishes within the Bills of Mortality.*

Every master bricklayer to give twenty-four hours notice to the Surveyor of the district from the first to the seventh rate, concerning the building to be altered or erected; but if the building is to be piled or planked, or begun with wood, it becomes the business of the carpenter to give such notice.

The footings of the walls are to have equal projections on each side: but where any adjoining building will not admit of such

projection to be made on the side adjoining to such building, to be done as near as the case will admit according to each of the four rates.

The act calls every front, side, or end wall, &c. (not being a party-wall) an external wall.

The timbers in each rate may be supposed to be girders, beams, or trimming joists, &c. and their bearing in all cases, and in all the above four rates, may be as much as the nature of the wall will admit, provided there is left four inches between the ends of such timber and the external surface of the wall.

The joints of the brick work may also be shown, and may answer to the express number of bricks of which such wall is to be composed.

It may now be necessary here to say something farther relative to external walls.

External Walls,

And other external enclosures, to the first, second, third, fourth, and fifth rate of building, must be of brick, stone, artificial stone, lead, copper, tin, slate, tile, or iron; or of brick, stone, artificial stone, lead, copper, tin, slate, tile, and iron together, except the planking, piling, &c. for the foundation, which may be of wood of any sort.

If any part to an external wall of the first and second rate, is built wholly of stone, it is not to be less in thickness than as follows:

First rate, fourteen inches below the ground floor, nine inches above the ground floor; second rate nine inches above the ground floor.

Where a recess is meant to be made in an external wall, it must be arched over, and in such a manner, as that the arch and the back of such recess shall respectively be of the thickness of one brick in length: it is therefore plain, that where a wall is not more than one brick thick, it cannot have any recess.

No external wall to the first, second, third, and fourth rate, is

ever to become a party wall, unless the same shall be of the height
and thickness above the footing, as is required for each party-wall
to its respective rate.

Of Party Walls.

Buildings of the first, second, third, and fourth rate, which are
not yet designed by the owner thereof to have separate and distinct side walls, on such parts as may be contiguous to other
buildings, must have party-walls; and they are to be placed half
and half on the ground of each owner, or of each building respectively, and may be built thereon, without any notice being
given to the owner of the other part; that is to say, the first builder
has a right so to do, where he is building against vacant ground.

Party-walls, chimnies, and chimney shafts hereafter to be built,
must be of good sound bricks or stone, or of sound bricks and
stone together, and must be coped with stone, tile, or brick.

Party-walls or additions thereto, must be carried up thirteen
inches above the roof, measuring at right angles with the back of
the rafter, and twelve inches above the gutter of the highest
building which gables against it; but where the height of a party-
wall so carried up, exceeds the height of the blocking course or
parapet, it may be made less than one foot above the gutter, for
the distance of two feet six inches from the front of the blocking
course or parapet.

Where dormers or other erections are fixed in any flat or roof,
within four feet of any party-wall, such party-wall is to be carried
up against such dormer, and must extend at least two feet wider,
and to the full height of every such dormer or erection.

No recess is to be hereafter made in any party-wall of the first,
second, third, and fourth rate, except for chimney-flues, girders,
&c. and for the ends of walls or piers, so as to reduce such wall
in any part of it to a less thickness than is required by the act,
for the highest rate of building to which such wall belongs.

No opening is to be made in any party-wall except for communication from one stack of warehouses to another, and from one

staple building to another, all which communications must have wrought-iron doors, and the panels thereof are not to be less than a quarter of an inch thick, and to be fixed in stone door-cases and sills. But there may be openings for passages or ways on the ground, for foot passengers, cattle, or carriages, which must be arched over throughout with brick or stone, or brick and stone together, of the thickness of a brick and a half at the least, to the first and second rate, and one brick to the third and fourth rate. And if there is any cellar or vacuity under such passage, it is to be arched over throughout in the same manner as the passage over it.

No party wall or party arch, or shaft of any chimney, new or old, must be cut into, other than for the purposes as follows :

If the fronts of buildings are in a line with each other, a recess may be cut, both in the fore and back fronts of such buildings, (as may be already erected) for the purpose of inserting the end of such other external wall, which is to adjoin thereto, this recess must not be more than nine inches deep from the outward faces of such external walls, and to be cut beyond the centre of the party-wall thereto belonging.

And further, for the use of inserting bressummers and story posts, that are to be fixed on the ground floor, either in the front or back wall, the recess may be cut from the foundation of such new wall to the top of such bressummer, fourteen inches deep from the outward face of such wall, and four inches wide in the cellar story, and two inches wide on the ground story.

And further, for the purpose of tailing-in stone steps, or stone landings, as for bearers to wood stairs, or for laying-in stone corbels for the support of chimney jambs, girders, beams, purlins, binding or trimming joists, or other principal timbers.

Perpendicular recesses may also be cut in any party-wall, whose thickness is not less than thirteen inches, for the purpose of inserting walls and piers therein, but they must not be wider than fifteen inches, or more than four inches deep, and no such recess is to be nearer than ten feet to any other recess.

No. 14.   2 c

All such cuttings and recesses must be immediately made good, and effectually pinned up, with brick, stone, slate, tile, shell, or iron, bedded in mortar.

No party-wall to be cut for any of the above purposes, if the same will injure, displace, or endanger the timbers, chimnies, flues, or internal finishings of the adjoining buildings.

The act also allows the footing to be cut off on the side of any party-wall, where an independent side-wall is intended to be built against such party-wall.

When any buildings (inns of court excepted) that are erected over gate-ways or public passages, or have different rooms and floors, the property of different owners, come to be rebuilt, they must have a party-wall, with a party arch or arches of the thickness of a brick and a half at least, to the first and second rate, and of one brick to the third and fourth rate, between building and building, or between the different rooms and floors, that are the property of different owners.

All inns of court are excepted from the regulation as above, and are only necessitated to have party-walls, where any room or chamber communicates to each separate and distinct stair-case, and which are also subject to the same regulations as respect other party-walls.

If a building of a lower rate is situated adjoining to a building of a higher rate, and any addition is intended to be made thereto. the party-wall must be built in a such a manner. as is required for the rate of such higher rate of building as adjoining.

When any party-wall is raised, it is to be made the same thickness as the wall is of, in the story next below the roof of the highest building adjoining, but it must not be raised at all, unless it can be done with safety to such wall, and the building adjoining thereto.

Every dwelling house to be built, which contains four stories in height from the foundation, exclusive of rooms in the roof, must have its party-wall built according to the third rate, although such dwelling-house may be of the fourth rate.

And every dwelling-house to be built in future which exceeds four stories in height, from the foundations, exclusive of the rooms in the roof, must have its party-wall built according to the first rate, although such house may not be of the first rate.

Chimnies, &c.

No chimney is to be erected on timber, except on the piling, planking, &c. of the foundations of building.

Chimnies may be built back to back in party-walls; but in that case, they must not be less in thickness from the centre of such party-wall than as follows:

First rate, or adjoining thereto, must be one brick thick in the cellar story, and half a brick in all the other stories.

Second, third, and fourth rate, or adjoining thereto, must be three-quarters of a brick thick in the cellar story, and half a brick in all the upper stories.

Such chimnies in party-walls as do not stand back to back, may be built in any of the four rates as follows:

From the external face of the party-wall to the inward face of the back of the chimney in the cellar story, one brick and a half thick, and in the upper stories, one brick thick from the hearth to twelve inches above the mantel.

Those backs of chimnies which are not in party-walls to the first rate, must not be less than a brick and a half thick in the cellar story, and one brick thick in every other story, and to be from the hearth to twelve inches above the mantel.

If such chimney is built against any other wall, the back may be half a brick thinner than that which is above described.

Those backs of chimnies which are not in party-walls of the second, third, and fourth rate, must be in every story one brick thick at least, from the hearth to twelve inches above the mantel.

These backs may be also half a brick thinner, if such chimney is built against any other wall.

All breasts of chimnies, whether they are in party-walls or not, are not to be less than one brick thick in the cellar story, and half a brick thick in every other story.

All withs between flues must not be less than half a brick thick.

Flues may be built opposite to each other in party-walls, but they must not approach to the centre of such wall nearer than two inches.

All chimney breasts next to the rooms, and chimney backs also, and all flues, are to be rendered or pargetted.

Backs of chimnies and flues in party-walls against vacant ground, must be lime whited, or marked in some durable manner, but must be rendered or pargetted as soon as any other building is erected to such wall.

No timber must be over the opening of any chimney for supporting the breast thereof, but must have a brick or stone arch, or iron bar or bars.

All chimnies must have slabs or foot paces of stone, marble, tile, or iron, at least eighteen inches broad, and at least one foot longer than the opening of the chimney when finished, and such slabs or foot paces must be laid on brick or stone trimmers at least eighteen inches broad from the face of the chimney breast, except there is no room or vacuity beneath, then they may be bedded on the ground.

Brick funnels must not be made on the outside of the first, second, third, or fourth rate, next to any street, square, court, road, or way, so as to extend beyond the general line of the buildings therein.

No funnel of tin, copper, iron, or other pipe for conveying smoke or steam, must hereafter be fixed near any public street, square, court, or way, to the first, second, third, or fourth rate, and no such pipe is to be fixed on the inside of any building nearer than fourteen inches to any timber, or other combustible material whatever.

INDEX

AND

EXPLANATION OF TERMS

USED IN

BRICKLAYING.

N. B. *This Mark § refers to the preceding Sections, according to the Number.*

———◆———

A.

ACT, BUILDING, page 206.
ARCHES, § 37. *See Plate* XXV.
ARRIS WAYS, tiles laid diagonally.
AXIS OF A VAULT, § 36.

B.

BANKER, § 19, 37. *See Plate* XXIII. *Fig.* 13.
BED OF A BRICK, the horizontal surfaces as disposed in a wall.
BEDDING STONE, § 22.
BEVEL, § 24, 37. *See Plate* XXIII. *Fig.* 4.
BOND, § 35.
BONE ASHES, § 32.
BORER, § 34.
BOSS, a short trough for holding water, when tiling the roof; it is hung to the lath.
BRICK AXE, § 28, 37. *See Plate* XXIII. *Fig.* 2.

Q 2

BRICK TRIMMER, a brick arch abutting upon the wooden trimmer under the slab of the fire place, to prevent the communication of fire.

BRICK TROWEL, § 4, 37. *See Plate* **XXIII**. *Fig*. 1.

BRICKLAYING, § 1.

BRICKS, § 33.

BUILDING ACT, § 38.

C.

CAMBER SLIP, § 20, 37. *See Plate* **XXIII**. *Fig*. 12.

CEMENTS, § 32.

CENTERING TO GROINS. *See Plate* **XXVIII**.

CHOPPING BLOCK, § 30.

CLAMP, § 33.

CLINKERS, hard bricks imported from Holland, § 33.

CLOSER, a brick-bat inserted where the distance will not permit of a brick in length. *See Plate* **XXIV**.

COMPASS, § 12.

CONIC SURFACES, § 36.

CONO-CONIC ARCH, § 36.

CONO-CONIC GROIN, § 36.

COURSE, a horizontal row of bricks stretching the length of a wall.

CROSS PASSAGES, § 36.

CUTTING BRICKS, § 33.

CYLINDRIC GROIN, § 36.

CYLINDRO-CYLINDRIC ARCH, § 36.

CYLINDRO-CYLINDRIC GROIN, § 36.

CYLINDRO-SPHERIC ARCH, § 36.

CYLINDRO-SPHERIC GROIN, § 36.

CYLINDROID, § 36.

CYLINDROID GROIN, § 36

D.

DUTCH CLINKERS, § 33.

PLACE BRICKS, § 33.
PLUMB RULE, § 6.
POZZOLONA, § 32.

R.

RAKER, § 13, 37. *See Plate* XXIII. *Fig.* 8.
RAMMER, § 16, 37. *See Plate* XXIII. *Fig.* 10.
RECTANGULAR GROIN, § 36.
ROD, § 9.
RUBBING STONE, § 21, 37. *See Plate* XXIII. *Fig.* 13.

S.

SAIL-OVER, is the overhanging of one or more courses beyond the naked of the wall.
SAW, § 27.
SCRIBE, § 26.
SCURBAGE, § 3.
SIMPLE VAULT, § 36.
SKEW BACK, the sloping abutment for the arched head of a window.
SOMMERING, the continuation of the joints of arches towards a centre or meeting point.
SPHERIC GROIN, § 36.
SPHERIC VAULT, § 36.
SPHERIC-CYLINDRIC ARCH, § 36.
SPHERO-CYLINDRIC GROIN, § 36.
SPRINGING LINES, § 36.
SQUARE, § 23, 37. *See Plate* XXIII. *Fig.* 3.
STEENING WELLS. *See Plate* XXX.
STRAIGHT ARCHES, heads of apertures, which have a straight intrados in several pieces, with radiating joints or bricks tapering downwards.
STRAIGHT VAULTS, § 36.
STRETCHERS, § 35.
STRETCHING COURSES, § 35.

2 D

T

TEMPLET, § 29.

TIN SAW, § 27.

TOOTHING, bricks projecting at the end of a part of a wall, in order to bond a part of the said wall not yet carried up.

TRIMMER, *See Brick Trimmer*.

V.

VAULTING, § 36.

WALLS, § 35.

WATER CEMENTS, § 32.

WATER TABLE, bricks projecting below the naked of a wall, in order to rest the upper part firmly.

MASONRY.

§ 1. MASONRY is the art of preparing and combining stones by such a disposition as to tooth or indent them into each other, and form regular surfaces for shelter, convenience, and defence, as the habitation of men, animals, goods, fortifications, bridges, separation of property, &c. and may be said to consist either of walling or arching.

§ 2. MASONS' TOOLS

The tools employed by the mason, are different in different counties, according to the quality of the stone employed: in some counties of England the stone is soft, with so little grit as to be wrought by planes into mouldings, as in joinery work; the naked surfaces of a building are generally finished with an instrument called a drag: the Bath and Oxfordshire stone is of this description. In other parts, the stone is so hard as only to be wrought by a mallet and chisel. In London, the value of stone occasions it to be cut into slips and scantlings by a saw; the operation is done by a labourer. In those countries were stone abounds, it is divided into smaller scantlings, by means of wedges. In most descriptions of stone, whether hard or soft, a hammer is employed in knocking and axing off the prominent parts. Hard stone and marble are reduced to a surface by means of a mallet and chisel. In rough stone from the quarry, where the saw has not been employed, a narrow chisel, called a point, about a quarter of an inch at the entering part, is first used; but the inequalities of sawn stone, if not very prominent, are reduced by means of an inch chisel, and sometimes more or less, according to the quantity to be wrought off. Chisels are from a quarter of an inch to three inches in breadth, at the cutting part: those of the greatest breadth are called tools, and employed finally on the surface, which is more regular after having gone over it, than that left after the operation of a narrow chisel. When the surface is wrought into narrow furrows or channels at regular distances, like small flutings which completes the finish of the face, the operation is called tooling, and the surface itself is said to be tooled. When the surface is required to be smoothed, it is done by rubbing it with a flat stone of the same kind with sand and water, and the larger the stone the more regular will the surface be.

The form of masons' chisels is like that of a wedge, the cutting edge is the vertical angle; they are wholly constructed of iron, except the steel end, which enters the stone. The end which is struck by the mallet, is a flat portion of a spheric surface, and projects on all sides beyond the handling part, which tapers upwards with an equal concavity on each side. The other tools used by the mason are a level, a plumb rule, a square, a bevel, a trowel, a mallet, a hammer, and sometimes a pair of compasses. These have been sufficiently treated under the former departments of Carpentry and Bricklaying, to which the reader is referred. The saw, as has been observed, though an appendage of Ma sonry, is used by the labourer.

§ 3. Of Marbles and Stones.

Marble is polished by being first rubbed with grit stone, afterwards with pumice stone, and lastly with emery or calcined tin. Marbles with regard to their contexture and variegation of colour, are almost infinite; some are black, some white, and some of a dove colour; the best kind of white marble is called statuary, which when cut into thin slices, becomes almost transparent, which property the other kinds do not possess. Other species of marble are streaked with clouds and veins. The texture of marble is not altogether understood, even by the best workmen, but they generally know upon sight, whether it will receive a polish or not. Some marbles are easily wrought, some are very hard, other kinds resist the tools altogether. Artificial marble or scagliola, is real marble, pulverized and mixed with plaster, and is used for columns, baso relievos, and other ornaments.

The chief kind of stone used in London, is Portland stone, which comes from the island of Portland in Dorsetshire; it is used for buildings in general, as strings, window sills, balusters, steps, copings, &c.; but under great weight or pressure, it is apt to splinter, or flush at the joints. When it is recently quarried, it is soft and works easily, but acquires great hardness in length of time. St. Paul's Cathedral and Westminster Bridge, are constructed of Portland stone.

Purbeck stone comes from an island of the same name, also in Dorsetshire, and is mostly employed in rough work, as steps and paving.

Yorkshire stone is also used where strength and durability are requisites, as in paving and coping.

Ryegate stone is used for hearths, slabs, and covings.

Mortar is used by masons in cementing their works. This has already been fully handled under the bricklaying department, which the reader may consult. In setting marble or fine work, they use plaster of Paris; and in water works, tarras is employed.

Tarras is a coarse mortar, durable in water, and in most situations. Dutch tarras is made of a soft rock stone, found near Cologne on the Rhine. It is burnt like lime, and reduced to powder by mills, from thence carried to Holland, whence it has acquired the name of Dutch tarras. It is very dear, on account of the great demand there is for it in aquatic works.

An artificial tarras is formed of two parts of lime, and one of plaster of Paris; another sort consists of one part of lime, and two parts of well-sifted coal ashes.

§ 4. *Stone Walls*

Are those built of stone, with or without cement in their joints; the beading joints have most commonly a horizontal position in the face of the work, and this ought always to be the case when the top of the wall terminates in a horizontal plane or line: in bridge buildings, and in the masonry of fenced walls upon inclined surfaces, the beading joints on the face sometimes follow the direction of the top or terminating surface.

The footings of stone walls ought to be constructed of large stones, which if not naturally nearly square from the quarry should be reduced by the hammer to that form, and to an equal thickness in the same course, for if the beds of the stones of the foundation taper, the superstructure will be apt to give way, by resting upon mere angles or points, or upon inclined surfaces : the courses of the footing ought to be well bedded upon each other with mortar ; and all the upright joints of an upper course should break joint, that is, they should fall upon the solid part of the stones in the lower course, and not upon the joints.

The following are methods practised in laying the footings of a stone foundation : when the walls are thin, and stones can be go. conveniently, that their length may reach across each course, from one side of the wall to the other, the setting of each course with whole stones in the thickness of the wall, is to be preferred. But when the walls are thicker, and bond stones in part can only be conveniently procured, then every other succeeding stone in the course, may be a whole stone in the thickness of the wall ; and every other interval may consist of two stones in the breadth, that is, placing the header and stretcher alternately, like Flemish bond in nine inch brick work. But when bond stones cannot be had conveniently, every alternate stone should be in length two-thirds of the breadth of the footing upon the same side of the wall; then upon the other side of the wall a stone of one-third of the breadth of the footing, should be placed opposite to one of two-thirds, and one of two-thirds opposite to one of one-third : so that the stones may be placed in the same manner as those of the other side.

In broad foundations where the stones cannot be procured for a length equal to two-thirds of the foundation, then build them alternately, with the joints on the upper bed of each footing, so that the joint of every two stones may fall as nearly as possible in the middle of the length of one or of each adjoining stone, observing to dispose the stones on each side of every footing.

A wall which is built of unhewn stone, laid with or without mortar, is called a rubble wall: they are of two kinds, coursed and uncoursed; the most kind of rubble is the uncoursed, of which the greater part of the stones are crude as they came out of the quarry, and a little hammer dressed. This kind of walling is very inconvenient for the building of bond timbers, but if they are to be preserved to plugging, the backing must be levelled at every height in which the bond timbers are disposed.

The best kind of rubble is the coursed; the courses are all of accidental thicknesses, adjusted by a sizing rule, the stones are either hammer dressed or axed: this kind of work is favourable for the disposition of bond timbers ; but as all buildings constructed either in whole or in part of timber are liable to be burnt, strong well built walls should never be bonded with timber, but should rather be plugged, for if such accident takes place, the walls will be less liable to warp.

Walls faced with squared stones, hewn or rubbed, and backed with rubble, stone, or brick, are called ashlar: the medium size of each ashlar measures horizontally in the face of the wall about twenty-eight or thirty inches, in the altitude twelve inches, and in the thickness eight or nine inches. The best figure of stones for an ashlar facing are formed like truncated wedges, that is to say, they are thinner at one end than at the other in the thickness of the wall, though level on the beds; so that when the stones of one course, or part of a course, are shaped in this manner, and alike situated to each other, the backs of the course will form an indentation, like the teeth of a joiner's saw, but more shallow in proportion to the length of a tooth: the next course has its indentations found in the same way, and the stones so selected that the upright joints break upon the solid of the stones below. By these means the facing and backing are toothed together, and unquestionably stronger than if the back of each ashlar had been parallel to the front surface of the wall; as the stones are mostly raised in the quarries of various thicknesses, in an ashlar facing it would contribute greatly to the strength of the work, to select the stones in each course, so that every alternate ashlar may have broader beds than those of every ashlar placed in each alternate interval.

In every course of ashlar facing, bond stones should be introduced, and their number should be proportional to the length of the course; this should be strictly attended to in long ranges of stones, both in walls without apertures, and in the courses that form wide piers; when they are wide, every bond stone of one course should fall in the middle of every two bond stones in the course below. In every pier where the jambs are coursed with the other ashlar in front, and also in every pier where the jambs are one entire height, every alternate stone next to the aperture in the former case, and every alternate stone next to the jambs in the latter case, should bond through the wall, and also every other stone should be placed lengthways in each return of each angle, not less than the average length of an ashlar. Bond stones should have no taper in their beds, the end of every bond stone, as well as the end of every return stone, should never be less than a foot, there should be no such thing as a closer permitted, unless it bond through the wall. All the uprights or joints should be square, or at right angles to the front of the wall, and may recede about three-fourths of an inch from the face with a close joint from the face, with a close joint from thence, gradually widening to the back, and thereby make hollow wedge-formed figures, which will give sufficient cavities for the reception of packing and mortar. Both the upper and lower beds of every stone should be quite level, and not form acute angles as is often the case; the joints from the face to about three-fourths of an inch within the wall, should either be cemented with fine mortar, or with a mixture of oil, putty, and white lead : the former is the practice both in London and Edinburgh, and the latter in Glasgow. The putty cement will stand longer than most stones, and will remain prominent when the face of the stones has been corroded with age. The whole of the ashlar, except that mentioned of the joints towards the face of the wall, the rubble work and the core should be set and laid in the best mortar, and every stone should be laid on its natural bed. All wall-plates should be placed upon a number of bond stones, and particularly those of the roof where there are no tie-beams, by which means they may either be joggled upon the bonds, or fastened to them by iron and lead.

In building walls or insulated pillars of very short horizontal dimensions not exceeding the length of stones that can be easily procured, every stone should be quite level on the bed, without any degree of concavity, and should be one entire piece, between every two horizontal joints. This should be particularly attended to in piers, where the insisting weight is great, otherwise the stones will be in danger of splintering, and crushing to pieces, and perhaps occasion a total demolition of the fabric.

Vitruvius has left us an account of the manner of constructing the walls of the ancients, which was as follows : the Riticulated, is that wherein the joints run in parallel lines, making angles of forty degrees each, with the horizon in contrary ways, and consequently the faces of the stones form squares, of which one diagonal is horizontal, and the other vertical. This kind of wall was much used by the Romans in his time. The Incertain wall was formed of stones of which one direction of the joints was horizontal, and the other vertical : but the vertical joints of the alternate courses were not always arranged in the same straight line, all that they regarded was, to make them break joint : this manner of walling was used by the Romans antecedent to the time of Vitruvius, who directs that in both the reticulated and incertain walls, instead of filling the space between the sides with rubble promiscuously, they should be strengthened with abutments of

hewn stone or brick, or common flints, built in cross walls
two feet thick, and bound to the facing and backing with
cramps of iron. The Emplection consisted of two sides or shells
of squared stone, with alternate joints, and rubble core in the
middle.

The walls of the Greeks were of three kinds, named Isodomum,
Pseudisodomum and Emplection. The Isodomum had the courses
all of an equal thickness, and the other called Pseudisodomum
had the courses unequally thick ; in both these walls, whenever the squared work was continued, the interval or core was
filled up with common hard stones laid in the manner of bricks
with alternate joints. , The Emplection was constructed wholly
of squared stones, in these bond stones were placed at regular
intervals, and the stones in the intermediate distance were laid
with alternate joints in the same manner as those of the face
so that this manner of walling must have been much stronger than
the emplection of the Roman villages. This is a most strong
and durable manner of walling, and in modern times it may be
practised with the utmost success, but in the common run of
buildings it would be too expensive.

§ 5. *Stairs.*

When stairs are supported by a wall at both ends, nothing difficult can occur in the construction, in these the inner ends of the
steps may either terminate in a solid newel, or to be tailed into a
wall surrounding an open newel ; where elegance is not required,
and where the newel does not exceed two feet six inches. The
ends of the steps may be conveniently supported by a solid pillar,
but when the newel is thicker, a thin wall surrounding the newel
would be cheaper.

In the stairs of a basement story, where there are geometrical
stairs above, the steps next to the newel are generally supported
upon a dwarf wall.

§ 6. *Geometrical Stairs.*

Have the outer end fixed in the wall, and one of the edges of every step supported by the edge of the step below, and constructed with joggled joints, so that they cannot descend in the inclined direction of the plane, nor yet in a vertical direction, the sally of every joint forms an exterior obtuse angle, on the lower part of the upper step, called a back rebate, and that on the upper part of the lower step of course an interior one, and the joint formed of these sallies is called a joggle, which may be level from the face of the risers, to about one inch within the joint. Thus is the plane of the tread of each step continued one inch within the surface of each riser, the lower part of the joint is a narrow surface, perpendicular to the inclined direction or soffit of the stair at the end next to the newel.

In stairs constructed of most kinds of stone, the thickness of every step at the thinnest place of the end next to the newel, has no occasion to exceed two inches, for steps of four feet in length, that is, by measuring from the interior angle of every step perpendicular to the rake. The thickness of steps at the interior angle, should be proportioned to the length of the step; but allowing that the thickness of the steps at each interior angle is sufficient at two inches, then will the thickness of steps at the interior angles be half the number of inches that the length of the steps has in feet: thus a step five feet long, would be two inches and a half at that place.

The stone platforms of geometrical stairs, viz. the landings, half paces and quarter paces, are constructed of one, two, or several stones, according as they can be procured. When the platform consists of two or more stones, the first platform stone is laid upon the last step that is set, and one end tailed in and wedged into the wall; the next platform stone is joggled or rebated into one set, and the end also fixed into the wall, as that and the preceding steps are, and every stone in succession, till the platform is completed. If there is occasion for another flight of steps, the

last platform becomes a spring stone for the next step, the joint is
to be joggled as well as all the succeeding steps, in the same
manner as the first flight.

Geometrical stairs executed in stone depend upon the following
principle : that every body must at least be supported by three
points, placed out of a straight line ; and consequently, if two
edges of a body in different directions be secured to another body,
the two bodies will be immoveable in respect to each other. This
last is the case in a geometrical stair, one end of a stair stone is
always tailed into the wall, and one edge either rests on the ground
itself, or on the edge of the preceding stair stone, whether the
stair stone be a plat or step. The stones forming a platform, are
generally of the same thickness as those forming the steps.

§ 7. *A short Account of the Origin of the Arch, and Authors who
may be consulted.*

The arch is perhaps one of the most useful inventions that ever
took place in the art of building ; by it we are enabled to cross
the deepest rivers and valleys, and places which are rendered
impassable by rocks or precipitous banks. In such situations,
without its aid, goods conveyed by inland navigation, or by any
other means, could never obtain the same celerity of transportation,
nor have been conducted at so easy a rate of expense. By the
use of the arch we are enabled to build apartments secure from
fire, to cover apertures where it would be impossible to lintle
them with stone, and to support walls or their tops almost to any
height.

The theory of the equilibrium of arches depends on the deepest
principles of mathematical science. Those who are desirous of
obtaining the fundamental part of the art of building arches, will
do well to consult the fifth article of *Emerson's Miscellanies,* and
Hutton's and Gwilt's Principles of Arches, and for a knowledge of
the practice, it will be well to peruse a work in French, by *Per-*

ronet, which has gained him great reputation, as containing the whole result of his experience in the practice of building bridges and arches : also a work by *Semple*, containing many excellent practical remarks; there are other authors, but those here spoken of, have acquired the most celebrity.

Arches are to be found in the Greek theatres, Stadia and Gymnasia, some of them erected probably 400 years before the Christian era. The most ancient arches of which we have any thing like dates, are the Cloaca at Rome, begun by Tarquinius Priscus. The emperor Adrian threw a bridge over the Cephisus between the territories of Attica and Elusis, on the most frequented road of Greece. The ancient bridges at Rome were eight in number : the most considerable of which was the Pons Ælius, now the bridge of Santo Angelo. There are several other Roman bridges in and out of Italy, but the most celebrated was that erected over the Danube by the emperor Trajan, the span of the arches is supposed to have been 170 feet each : but even this is considerably surpassed in horizontal extent by the ancient bridge of Brioude in France, consisting only of one arch of 181 feet span. Several of the French bridges are remarkable for the great extent of the arches. The bridge of Neuilly, built by M. Perronet over the Seine, consists of five elliptic arches, each 128 feet span, composed of eleven arcs of circles, of different radii. The most considerable arch in Great Britain, is that over the river Taff, near Llantrissent in Glamorganshire, consisting of one arch of 140 feet span : the curve is the arc of a circle of 175 feet diameter. Sarah, or Island bridge over the Liffey, above Dublin, consists of one arch of 106 feet span. The bridges at Westminster, and Blackfriars, London, though among the boldest and finest undertakings of modern times, have their arches of less horizontal extension than those above mentioned ; the arches of the former are semi-circular, the central one is seventy-six feet diameter or span. The arches of the latter are nearly elliptic, nine in number, and the central one is 100 feet wide, and the arches on each side decrease regularly to the land piers.

n 2

PLATE XXXI.

Observations on the customary Problems in Masonry respecting Arches, and Methods of describing Elliptic Arches.

The operation of describing an ellipse with a string, though true in principle, is useless in practice, as the string stretches in such a degree as not to be depended on, and the degree of tension is in proportion to the length of the string, which is therefore unfit to be used for describing the curve of an arch of large extent. The trammel or elliptic compass is a very accurate instrument, but it can only be used for works upon a small scale: this method of description will be found in Problem V. Geometry. The description of an ellipse with a beam compass may be put in execution in arches of any extent, as has been fully verified in the practice of that distinguished French engineer, M. Perronet. But the common method with three centres only is extremely lame, owing to the sudden variation of curvature, which takes place at the junction of two very different radii.

———

PROB. I. *To render the Compass Method useful, not only in describing the Curve, but in finding the Joints perpendicular thereto, so as to form an Arch which shall not have any sensible variation in Practice from the true Elliptic Curve, nor in the Perpendicularity of the Joints.*

Find a number of points in the curve equidistant on each side of the extremity of the conjugate axis: find the centre of a circle passing the middle point, and the other two points one on each side of it: join the centre with the last two points of the curve, and describe an arc through the three points; then to complete the half curve, join one of the next points of the curve and the end of the arc by a straight line; or suppose these two points to be joined, and

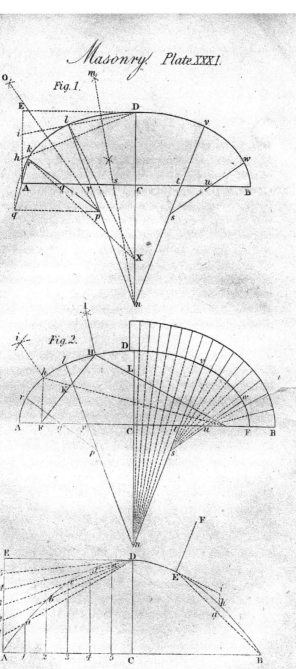

Masonry. Plate XXXI.

Fig. 1.

Fig. 2.

bisect this line by a perpendicular, which produce until it meet the first of the radii : join the last point of the curve, and the concourse of the two last radii : from the point of concourse describe an arc from the end of the arc last described to the next point in the curve ; proceed in like manner with the next succeeding arcs, if more than two, until the last arc but one, is described : continue the last arc until it meet a diameter parallel to the transverse axis : draw a line from the meeting of the arc and diameter through the extremity of the transverse axis, and produce this line till it meets the arc ; from the point where the line meets the arc draw a line to the centre of the arc ; from the point where the line so drawn cuts the transverse axis as a centre, describe an arc from the end of the arc last described to the extremity of the transverse axis.

Example, Fig. 1. Let A B be the transverse axis, and C D the semi-conjugate.

Draw E D parallel to A C and A E parallel to C D. Divide C A and A E each into three equal parts at the points *f, g, h, i.* Produce D C to X making C X equal to C D. Draw X *f l* and *x g k,* also *h k d* and *i l d,* then the points *k* and *l* will be in the curve, bisect the distance *l* D at right angles by *m n* meeting D X produced at *n*. Join *l n* cutting A C at *y*. The points *t* and *u* being on the line or semi-transverse C B, make C *t* equal to C *y,* and draw *n t v*. From *n* with the distance *n* D or *n l,* describe the arc *l* D *v*. Bisect the distance *k l* by a perpendicular *o p* meeting *l n* at *p*. From *p* with the distance *p l* describe the arc *l k q*. Draw *p q* parallel to A B. Join *q* A which produce to meet the arc *l k q* in *r* ; also join *r p* cutting A B in *g*. From *g* with the distance *g r* describe an arc *r* A, and the half A D and part of the other half D *v* of the arch will be completed. Make *t u* equal to *f g, n s* equal to *n p*. Draw *s u w*. From *s* describe the arc *v w*, and from *u* describe the arc *w* B which will complete the other half of the arch.

Prob. ii. *To find the Joints of an Elliptic Arch at right Angles to*
the Curve.

Fig. 2. Find the centres *n, p, s, g, y, t, u* as in Problem 1.,
then radiate the joints between D and *v* by the centre *n*, the joints
between *v* and *w* by the centre *s*, and the joints between *w* and B
by the centre *u*, and the other half of the arch A D in the same
manner, or thus:

If the arch A D B is described with a trammel, Take the semi-
transverse A C, and from D describe an arc cutting C A at F,
and another cutting C B at F, then the points F, F are called the
focii. Now to draw a line at right angles to the curve from any
point H. Draw H K F and H L F, making H K equal to H L.
From K and L as centres, describe arcs of equal radii cutting
each other at I, and draw I H, which will be a joint at the required
point H. In the same manner may any other joint *i h* or as many
as required be obtained.

———

Prob. iii. *To describe the Parabolic Arch, and thence to draw the*
Joints at right Angles to the Curve.

First, to draw the Curve.

Fig. 3. Let C D be the abscissa or height of the curve, and
E B the base or a double ordinate. Draw A E parallel to C D
and E D parallel to A B. Divide C A and A E each into the
like number of equal parts. Draw *i a*, 2 *b*, 3 *c*, &c. parallel to
C D; also draw 1 D, 2 D, 3 D, &c. cutting the parallels at *a, b,*
c, &c. which are points in the curve, then the curve may be
drawn with a bent rule through the points, *a, b, c*, &c. and the
other half B D being drawn in like manner will complete the
whole curve.

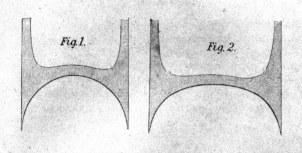

Fig.1.   Fig.2.

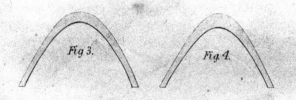

Fig 3.   Fig.4.

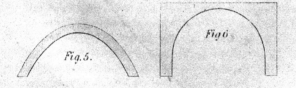

Fig.5.   Fig 6

Secondly, to find the Joints.

Let it be required to find a joint to any point E. Join E B. which bisect at *g :* draw *g h i* perpendicular to A B cutting the curve at *h :* make *h i* equal to *h g,* and join E *i :* draw E F at a right angle with E *i* and E F will be a joint at right angles to the curve. In the same manner all other joints may be obtained.

PLATE XXXII.

With respect to the power which arches have of supporting themselves, it depends upon the load insisting on all points of the arch, it is evident that there may be such a relation between the curve and the weight on every point of it, so as the weight may have on more tendency to break or spring the arch in one point than another; and it is evident, that if the materials are of the same specific gravity, that the wall erected at a given height upon the arch will obtain a certain form, so as to keep the arch in equilibrio, and that the form of the terminating line of the wall will depend on the curve of the supporting arch.

Fig. 1. If the intrados of the arch be a semi-circle, or semi-elliptic, the extrados or terminating line of the wall will be a curve running upwards at the ends, so as to make the two vertical lines which are tangents at the extremes of the arch asmytotes of the curve, and consequently, neither the semi-circular nor semi-elliptic arch are adapted to bridge building; and it may be pronounced with safety, that though these curves are frequently employed in bridge building, were the materials only placed in contact without cohesion or friction, the mass supported could not stand when the road way is straight, or a convex curve throughout the length of the arch, and that it is only in consequence of friction or the cementing quality of the mortar in connecting the whole of the materials in one mass, that such arches stand for so many centuries as they are found to do. However, by employing only the

2 F

middle portions of these curves, a road way or extrados of tolera-
ble convenient form may be obtained.

Fig. 3. Is an arch of equilibration, the intrados of which is
parabolic, which requires an extrados of the same form and
curvature, both being similar and equal. The vertical heights
between the two are everywhere equal.

Fig. 4 is another equilibrated arch, the intrados is an hyperbolio
curve, and the extrados requires a curve, such that the vertical
lines between the two curves are continually less from the crown
towards the feet of the arches.

Fig. 5 is another equilibrated arch, the intrados being a cate-
narian, or such as would be formed with a heavy chain suspended
at its extremities from two points at less distance from each other
than the length of the chain; the extrados to this curve may admit
of different forms, it may either be a convex curve, as when the
wall erected upon it is low, or a straight surface or plane, as when
the wall erected on it is enormously high, or a concave curve,
as when the wall is still higher: neither of the three last curves
are at all adapted to bridge building, the extrados line at a moderate height of wall being too rapid in its acclivity and declivity.

Fig. 6 is an arch of equilibration, where the top is a straight
line: the intrados at a given height of wall is calculated to answer
thereto, this arch is therefore well adapted in most situations for
the arch of a bridge.

INDEX

EXPLANATION OF TERMS

USED IN

MASONRY.

N. B. *This Mark § refers to the preceding Sections, according to the Number.*

A.

ABUTMENTS OF A BRIDGE, the walls adjoining to the land, which support the ends of the extreme arches or road way.

ARCH, in masonry, is a part of a building suspended over a hollow and concave towards the area of the hollow ; the top of the wall or walls which receives the first arch stones is called the abutment or springing, § 7.

ARCHIVOLT OF THE ARCH OF A BRIDGE, is the curve line formed by the upper sides of the arch stones in the face of the work, or the archivolt is sometimes understood to be the whole set of arch stones that appear in the face of the work.

ASHLAR, § 4.

B.

BANQUET, the raised footways adjoining to the parapet on the sides of a bridge.

BATH STONE, § 3.

BATTER, the leaning back of the upper part of the face of a wall, so as to make the plumb line fall within the base.

BATTARDEAU, OR COFFERDAM, a case of piling without a bottom, for building the piers of a bridge.

BEDS OF A STONE, are the parallel surfaces which intersect the face of the work in lines parallel to the horizon, § 4.

BOND, is that connection of lapping the stones upon one another in the carrying up of the work, so as to form an inseparable mass of building.

BOND STONES, stones running through the thickness of the wall in order to bind it.

BOND TIMBERS, § 4.

BRIDGE BUILDINGS, § 4.

BRIDGE, in masonry, is an edifice or structure, consisting of one or a series of arches, in order to form a road way over a river canal, &c. for passing the same.

BUTMENTS. *See Abutments.*

C.

CAISSON, a chest or box in which the piers of a bridge are built by sinking it as the work advances till it comes in contact with the bed of the river, and then the sides are disengaged, being constructed for the purpose.

CENTRES, the frames of timber work for supporting arches during their erection.

CHEST, the same as Caisson.

CHISELS, § 2.

COFFERDAM, the same as Battardeau

D.

DRAG, a thin plate of steel indented on the edge, like the teeth of a saw, used in soft stone which has no grit, for finishing the surface. A piece of a joiner's hand-saw makes a good drag, § 2.

DRIFT, the horizontal force of an arch, by which it endeavours to overset the piers.

DUTCH TARRAS, § 3.

E.

EMPLECTION, § 4.

ENTRADOS OF AN ARCH, the exterior or convex curve or the top of the arch stones : the term is opposed to the intrados or concave side.

EXTRADOS OF A BRIDGE, the curve of the road way.

F.

FENCE WALLS, those used to prevent the encroachments of men or animals.

FIGURES OF STONES, § 4.

FOOTINGS, projecting courses of stone without the naked of the superincumbent part, in order to rest the wall firmly on its base, § 4.

G.

GEOMETRICAL STAIRS, § 6.

H.

HEADERS, stones disposed with their length horizontally in the thickness of the wall.

I.

IMPOST OR SPRINGING, the upper part or parts of a wall in order to spring an arch.

INCERTAIN, § 4.

INSULATED PILLARS, § 4.

ISODOMUM, § 4.

J.

JETTE, the border made round the stilts under a pier.

JOGGLED JOINTS, the method of indenting the stones, so as to prevent the one from being pushed away from the other by lateral force, § 6.

K.

KEY STONE OF AN ARCH, the stone at the summit of the arch, put in last of all, for wedging and closing the arch.

KEY-STONE, the middle voussoir of an arch over the centre.

KEY-STONES, used in some places for bond stones.

L.

LEVEL, horizontal or parallel to the horizon.

LEVEL, an instrument, the same as that used in bricklaying and carpentry.

M.

MALLET, the implement or tool which gives percussive force to the chisel; in figure it approaches to a hemisphere, with a handle projecting from the middle or pole of the convex side, § 2.

MARBLE, § 3.

MASONRY, § 1.

MORTAR. *See Bricklaying*, § 32. *and in Masonry*, § 3.

N.

NAKED OF A WALL, is the vertical or battering surface, whence all projectures arise.

O.

OFF SET, the upper surface of a lower part of a wall left by reducing the thickness of the superincumbent part upon one side or the other, or both.

OXFORDSHIRE STONE, § 2.

P.

PARAPETS, the breast walls erected on the sides of the extrados of the bridge for preventing passengers from falling over.

PAVING, a floor or surface of stone for walking upon.

PIERS, the insulated parts of a bridge between the apertures, or arches for supporting the arches and road way.

PIERS IN HOUSES, the walls between apertures, or between an aperture and the corner.

PILES, timbers driven into the bed of a river, or the foundation of a building for supporting a structure.

PLASTER OF PARIS, § 3.

PITCH OF AN ARCH, the height from the springing to the summit of the arch.

POINT, the narrowest of all the chisels, and used in reducing the rough prominent parts of stone, § 2.

PORTLAND STONE, § 3.

PSEUDISODOMUM, § 4.

PURBECK STONE, § 3.

PUSH OF AN ARCH, the same as Drift, which see.

Q.

QUARRY, the place whence stones are raised.

R.

RANDOM COURSES IN PAVING, unequal courses without any regard to equi-distant joints.

RETICULATED WALL, § 4.

RUBBLE WALL, § 4.

RYEGATE STONE, § 3.

S.

SAW, a thin plate of iron of considerable length, regulated by a frame of wood and cording : the operation is performed by the labourer, § 2.

SHOOT OF AN ARCH, the same as Drift or Push. *See Drift.*

STATUARY, § 3.

STERLINGS, a case made about a pier of stilts in order to secure it.

STILTS, a set of piles driven into the bed of a river, at a small distance from each other, with a surrounding case of piling driven closely together ; the tops of the piles being levelled to low water mark, and the interstices filled with stones, forms a foundation for building the pier uoon.

STONE STAIRS, § 5.

STONE WALLS, § 4.

STRETCHERS, those ctones which have their length disposed horizontally in the length of the wall.

T.

TARRAS, § 3.

THROUGH STONES, the term used in some counties for bond stones, which see.

THRUST, the same as Push, Shoot, or Drift. *See Drift.*

TOOLING, § 2.

TOOLS, § 2.

U.

UNDER BED OF A STONE, the lower surface generally horizontally posited.

UPPER BED OF A STONE, the upper surface generally horizontally posited.

V.

VAULT, a mass of stones so combined as to support each other over a hollow.

Voussors the arch stone in the face or faces of an arch, the middle one is called the key-stone.

W.

Wall, an erection of stone generally perpendicular to the horizon and sometimes battering, in order to give stability.

Y

Yorkshire Stone, § 3.

S L A T I N G.

§ 1. SLATING is the operation of covering the top or other inclined parts of a building with slate.

SLATERS' TOOLS

Are a scantle, a trowel, a hammer, a zax, a small hand pick, a hod, a board for mortar. See the following explanation of terms.

EXPLANATION OF TERMS

USED IN

SLATING.

———

B.

BACK OF A SLATE, is the upper side of it.

BACKER, is a narrow slate put on the back of a broad square-headed slate, when the slates begin to get narrow.

BED OF A SLATE, is the lower side.

BOND OR LAP OF A SLATE, is the distance between the nail of the under slate, and the lower end of the upper slate.

C.

COURSE, is any row of slating, the lower ends of which are horizontally posited.

E.

EAVE, the skirt or lower part of the slating hanging over the naked of the wall.

H.

HOLING, the piercing of the slates for nails.

L.

Lap. *See Bond.*

M.

Margin of a Course, those parts of the backs of the slates exposed to the weather.

N.

Nails, painted iron or copper of a pyramidal form for fastening the slates to the lath or boarding

P.

Patent Slating, large slates used without boarding, and screwed to the rafters with slips of slates bedded in putty to cover the joints.

S.

Scantle, is a gauge by which slates are regulated to their proper length.

Slates used in London are of several kinds, as Westmoreland, rags, imperial, dutchess, countess, ladies, doubles. The Westmoreland are the best; they are from three feet six inches, to one foot in length, and from two feet six inches to one foot broad. Rags are the second best, and run nearly of the same size. The third in order, of inferior quality, are the imperials, they run from two feet six inches long, to one foot long. The other kinds will be understood by the order under which they are named, being inferior in size accordingly.

Sorting is the regulating of slates to their proper length by means of the scantle.

Squaring, the cutting of the sides and bottom of the slates.

T.

TAIL, the bottom or lower end of the slate.

TRIMMING, the cutting or paring of the side and bottom edges, the head of the slate never being cut.

Z.

Zax, the tool for cutting the slate.

PLASTERING.

§ 1. PLASTERING is the art of covering walls or ceilings with one, two, or three layers of any plastic or tenacious paste, so as to admit of a smooth and hard surface when the material is dry, and also of ornamenting walls and ceilings, either by being run or cast into moulds.

§ 2. PLASTERERS' TOOLS.

Tools used by the plasterer, are plastering trowels of several descriptions, joint trowels, and jointing rules, a hawke, a hand float, a quirk float, and a derby. A scratcher, and wooden skreeds for running mouldings.

§ 3. MATERIALS

Generally employed are lime, hair, sand, plaster of Paris; and these are variously compounded, as the following alphabetical arrangement of terms will show, which also explains the tools and their uses.

Walls consisting of brick or stone in the best houses, are always lathed by the plasterer, previous to the operation of plastering, particularly interior walls; and it is more requisite to lath walls constructed of stone, than those constructed of brick, which is a dry substance, and not liable to attract damps.

Ceilings are generally plastered upon laths, particularly in London. In some parts of the country, reeds are employed in their stead: the reeds are spread out on the ceiling, so as to form a regular surface, and are confined to their situation by nailing laths to the joists, the reeds running transversely between them and the joists. The reeds are cheaper than laths, but require more material of plaster and labour: so that when finished, the difference of cost is very trifling. Other matters in plastering will be seen in the following explanation of terms

EXPLANATION OF TERMS

IN

PLASTERING.

———◆———

A.

ANGLE FLOAT, is a float made to any internal angle to the planes of both sides of the room.

B.

BASTARD STUCCO, is three coat plaster, the first generally rough-ing-in or rendering, the second doating is in troweled stucco, out the finishing coat contains a little hair besides the sand: it is not hand floated, and the troweling is done with less labour than what is denominated troweled stucco.

BAY, a strip or rib of plaster between skreeds for regulating the floating rule.

C.

CEILING, the upper side of an apartment opposite to the floor, generally finished with plastered work. Ceilings are set in two different ways, the best is where the setting coat is composed of plaster and putty, commonly called gauge. Common ceilings have plaster but no hair, this last is the same as the finishing coat in walls set for paper

Coarse Stuff. *See Lime and Hair.*

Coat, a stratum or thickness of plaster work, done at one time

D.

Derby, a two-handed float.

Die, is when plaster loses its strength,

Dots, patches of plaster put on to regulate the floating rule in making skreeds and bays.

Double Fir Laths, are laths three-eighths of an inch thick, single fir laths being bare by a quarter. All the ceilings on the entrance and drawing-room floors and best stair-cases should be lathed with double fir laths.

F.

Fine Stuff is made of lime slacked and sifted through a fine sieve, and mixed with a due quantity of hair, and sometimes a small quantity of fine sand. Fine stuff is used in common ceilings and walls, set for paper or colour.

Finishing, is the best coat of three coat work, when done for stucco. The term setting is commonly used, when the third coat is made of fine stuff for paper.

First Coat of two coat work is denominated laying, when on lath, and rendering on brick, in three coat work upon lath it is denominated pricking-up, and upon brick, roughing-in.

Float, an implement for forming the second coat of three coat work to a given form of surface. Floats are of three kinds: namely, the hand float, the quirk float, and the derby.

Floated Lath and Plaster set fair for paper, is three coat work, the first pricking-up, the second floating, and the third or setting coat of fine stuff, understood to be pricked-up, as there is no floated work without pricking-up.

Floated, rendered and set, this is the common term.

Floated Work, is that which is pricked-up, floated and set, or roughed-in. 2 H

Floating, is the second coat of three coat work. There is no floating without pricking-up or roughing-in first, and then the finishing or setting. Floating consists of the same stuff as pricking-up, but more hair is used in the former than in the latter. The floating should be brushed with a birch broom, in order to rough the surface, for stucco or setting for paper. Floating is always used in stuccoed work, walls prepared for paper, and in the best ceilings.

Floating Skreeds differ from cornice skreeds in this, that the former is a strip of plaster, and the latter wooden rules for running the cornice.

Floating Rules are of every size and length.

G.

Gauge, a mixture of fine stuff and plaster, or putty and plaster, or coarse stuff and plaster, used in finishing the best ceilings and for mouldings, and sometimes for setting walls.

H.

Hair used in plastering, ought to be long fresh hair.

Hawke, a board with a handle projecting perpendicularly from the under side for holding the plaster.

J.

Joint-Rules and Tools are narrow trowels and rules of wood for making good mitres.

L.

Lath Floated and Set Fair. These words bear the same meaning as lath pricked-up and floated and set, which see.

Lath Layed and Set, is two coat work, only the first coat called laying, is put on without scratching, except it is swept with a broom. This is generally coloured on walls, and whited on ceilings.

LATH PLASTERED SET AND COLOURED, is the same with lath layed set and coloured, which see.

LATH PRICKED-UP FLOATED AND SET FOR PAPER, is three coat work, the first is pricking-up, the second floating, and the finishing is fine stuff.

LAYING, is the first coat on lath of two coat plaster or set work it is not scratched with the scratcher, but its surface is roughed by sweeping it with a broom. It differs only from rendering on its application. Rendering is applied to the first coat work upon brick, whereas laying is the first of two coat work upon lath.

LAYING-ON TROWELS, the trowels used for laying on the plaster.

LIME AND HAIR, is a mixture of lime and hair used in first coating and floating. It is otherwise denominated coarse stuff: in floating more hair is used than in first coating.

M.

MATERIALS in plastering are, coarse stuff, fine stuff, stuff, putty, plaster, gauge, and stucco.

MITERING ANGLES, in making good internal and external angles of mouldings.

MOULDINGS, when not very large, are first run with coarse gauge to the mould, then with fine stuff, then with putty and plaster, and lastly, run off or finished with raw putty. When mouldings are large, coarse stuff is first put on, then it is filled with tile heads or brick bats, and run off successively, with coarse gauge, fine stuff gauge, putty gauge, and finished with raw putty: in running cornices there must always be skreeds upon the ceiling, whether the ceiling is floated or not.

P.

PAIL, a vessel for holding water to moisten the plaster.

PLASTER, is the material with which ornaments are cast, and with which the fine stuff of gauge for mouldings and other parts are mixed.

PRICKING-UP is the first coating of three coat work upon laths. The material used is coarse stuff, and sometimes mixed up in London with road dirt or Thames sand, and its surface is always scratched with the scratcher.

PUGGING, the stuff laid upon sound boarding, in order to prevent the transmission of, or deaden the sound in its passage from one story to another.

PUTTY, is a very fine cement made of lime only. It is thus prepared: dissolve in a small quantity of water, as two or three gallons, so much fresh lime, (constantly stirred with a stick,) until the lime be entirely slacked, and the whole becomes of the consistency of mud; so that when the stick is taken out of it, it will but just drop; then being sifted or run through a hair sieve, to take out the gross parts of the lime, it is fit for use. Putty differs from fine stuff in the manner of preparing it, and in its being used without hair.

Q.

QUIRK FLOAT.   *See Angle Float.*

R.

RENDERED AND FLOATED, is three coat work, more commonly called floated, rendered and set.

RENDERED FLOATED AND SET, for paper, should be termed roughed-in; floated and set for paper is three coat work, the first lime and hair upon brick work, the second the same stuff with a little more hair floated with a long rule, the last fine stuff mixed with white hair.

RENDERED AND SET, the same as set work, *see Set Work.* Rendering is the first of two coat work upon naked brick or stone work whited on walls or vaults: roughing-in being the first coat of three work on naked brick, but the compound term pricking-up, is used for the first of three coat work upon lath, or on brick work, which has been previously rendered. Though the term rendering is sometimes used in three coat work, it is improper. The material for rendering is the same as that for pricking-up.

ROUGH CAST, is the overlaying of walls with mortar, without smoothing it with any tool whatever.

ROUGH RENDERING, is one coat rough.

ROUGH STUCCO, is that which is finished with stucco floated and brushed in a small degree with water, much used at present.

ROUGHING-IN, is the first coat of three coat work.

RUNNING MOULDINGS. *See Mouldings.*

S.

SCRATCHER, is the instrument for scratching the plaster, as its name implies.

SECOND COAT, is either the finishing coat, as in layed and set, or in rendered and set; or it is the floating, when the plaster is roughed-in floated and set for paper.

SET FAIR, is used after roughing-in and floated, or pricked up and floated: it should be well troweled, as it does not answer for colour without.

SET WORK, two coat work upon lath: the plasterers denominate set work by the compound term of layed and set.

SETTING COAT, on ceilings or walls in the best work, is gauge, or a mixture of putty and plaster; but in common work it consists of fine stuff, and when the work is very dry, a little sand is used. The setting coat may either be a second coat upon laying or rendering, or a third coat upon floating; the term finishing is applied to the third coat when of stucco, but setting for paper.

SETTING, is also the quality that any kind of stuff has to harden in a short time.

SINGLE FIR LATHS are something less than one fourth of an inch in thickness.

SKREEDS are wooden rules, for running mouldings. Skreeds are also the extreme guides upon the margins of walls and ceilings for floating, to the intermediate ones being called bays. In running cornices, where the ceilings are not floated, there must always be skreeds.

STOPPING, making good holes in the plaster.

STUCCO, OR FINISHING, is the third coat of three coat plaster, consisting of fine lime and sand; the best is twice hand floated and well troweled; bastard succo has a little hair. *See Finishing.* Rough stucco is only floated and brushed in a small degree with water: troweled stucco is accounted the best.

T.

TRAVERSING the skreed for cornices, is putting on gauge stuff on the ceiling skreeds, for regulating the running mould of the cornice above.

THREE COAT WORK, is that which consists of pricking-up or roughing-in, floating, and a finishing coat.

TROWELED STUCCO for paint, the same as roughed-in on brick work, and set or pricked-up, floated and twice hand floated.

THIRD COAT, is the stucco for paint or setting for paper.

TWO COAT WORK, is either layed and set, or rendered and set. *See these articles.*

W.

WALL, is the coating of plaster layed and set, and applied to brick work only where there are two coats.

PAINTING IN OIL.

Painting is the art of covering the surfaces of wood, iron, &c. with a mucilaginous substance, which shall acquire hardness on the surface, and thereby protect from the weather, and produce any colour proposed. It is intended here to treat only of common painting in oil, which comprehends the mechanical process for preserving and ornamenting stuccoed walls and wood work of houses; also iron and wooden rails, &c.

In this branch, the requisite tools are brushes of hogs' bristles of various sizes, suitable to the work, a scraping or pallet knife, earthen pots to hold the colours, a tin can for turpentine, a grinding stone and muller, &c.; the stone should be hard and close grained, about eighteen inches diameter, and sufficiently heavy to keep it steady.

The Process for Painting on new Wood Work.

As the *knots* in wood (particularly deal) are a great annoyance in painting, great care is required in what the painters term killing them, and the most sure way of doing this has been found to be, by laying upon those knots which retain any turpentine, a great substance of lime, immediately on its being slacked, with a stopping knife, (this process dries or burns up the turpentine which the knots contain,) and when the lime has remained on about twenty-four hours, scrape it off, then do them twice over with size knotting, which is made with red and white lead ground very fine with water on a stone, and mixed with strong double glue-size to be used warm, after which, if you have any doubts of their not being sufficiently covered, do them over with red and white lead ground very fine in linseed oil, and mixed with a proportion of that oil, taking care to rub them down with fine sand paper, each time you do them over, to prevent their appearing more raised than the other parts, by the repetition of a greater number of coats than the other parts of the work will have; when this is quite dry, lay on your *priming colour*, which is made with white and a little red lead mixed thin with linseed oil. When the priming is quite dry, and if the work is intended to be finished white, mix white lead, and a'very small portion of red with linseed oil, adding a very little spirits of turpentine, and second colour your work; it is well to let the work remain in this state for some days to harden: then your care must be (before you lay on your third coat) to rub it down with fine sand paper, and stop with oil putty wherever it may be necessary, observing particularly if any of the knots show through your work, in which case take silver leaf, and lay it upon them with japan gold size; the third coat is white lead mixed with linseed oil and turpentine in equal portions, and if the work is intended to be finished with four coats, let your finishing coat be made of good old white lead and thinned with bleached linseed oil and spirits of turpentine, of the portion of one of oil and two of turpentine; a very small quantity of blue black may be used in the two last coats; and if the work is to be flatted dead white, the above process is prepared to receive. *Dead white* is fine old Nottingham lead, and thinned entirely with spirits of turpentine.

In painting on *stucco*, it is necessary to give it one coat more than wood work, therefore the fourth coat should be mixed with half spirits of turpentine and half oil, and this will receive the finishing coat of all turpentine or flatting. But if not to be flatted, then the finishing coat should be done with one part oil and two of turpentine. As the colours used on stucco walls are very nume-rous, it would far exceed my limits to treat of them distinctly: let it therefore suffice to say, that the same process must be observed in using them as in white, only that each coat should incline to the colour they are intended to be finished.

———

The Process for Painting on old Work.

Let all the work you intend to paint be well rubbed down with dry pumice stone, and carefully dusted off, and where the work may require, let any cracks or openings be well stopped with oil putty, after which mix white lead, adding a very small portion of red lead, and with turpentine and oil of equal parts, paint your work (this coat is technically called by painters second colouring old work) after this is done and the work dry, mix good old white lead with half bleached oil and half turpentine, adding a very small portion of blue black, and finish your work: or if it is intended to be flatted, the former process is a proper preparation to receive the dead white; the same process is to be observed for stuccoed walls, observing, that if they require a great number of coats, the mixture of half oil and half turpentine is proper. The more you nix your colours with oil, and the less with turpentine for outside work the better, as turpentine is more adherent to water than oil, and consequently, not so well calculated to preserve work exposed to the weather; yet as oil will discolour white, it is necessary to finish that with a portion of half oil and half turpentine : but in dark colours, such as chocolate, greens, lead colour, &c. &c. boiled linseed oil and a little turpentine is the best, or boiled oil only.

White lead is used in all stone colours; white painting is entirely white lead ; lead colours are white lead and lamp black ; pinks and all fancy colours have a portion of white lead in their composition : but chocolates, black, brown, and wainscoats have no portion whatever.

Clear coaling is made of white lead ground in water and mixed

Nos. 17 & 18.　2 I

with size; it is used instead of a coat of paint, but by no means answers the end, as not possessing a sufficient body, and will scale off in time, and change the colour in damp situations. Clear coaling is most useful where the work is greasy and smoky, as it prepares it better to receive a coat of paint: but when used for joiners' work where mouldings are concerned, it destroys the accuracy of the workmanship by filling up the quirks and mitres of the mouldings. Clear coaling is not much used at present.

Some colours dry badly, and in damp weather all colours require something to expedite their drying; a *good dryer* may be prepared of equal parts of copperas and litharge ground very fine, to be added as wanted.

Putty is made of whiting and linseed oil, well beaten together.

The brushes when done with should be put into a pan with water, which prevents their drying and becoming hard; also if any colour is left, water should be put upon it to prevent its drying.

Drying oil is made thus: to every gallon of linseed oil put one pound of red lead, one pound of umber, and one pound of litharge. The oil and the materials to be boiled for two or three hours. *Note.* If the pot in which the oil is boiled will contain fifteen gallons, it is not prudent to boil more than five gallons at a time as the oil and material will swell so much as to endanger boiling over and setting the place on fire. After having boiled a sufficient time, the pot may then be filled up with oil, and made to simmer gently, and then it is finished.

A List of useful Colours for House Painting.

Black—lamp black.
White—white lead.
Yellow—ochers, also patent yellow.
Blue—Prussian blue, and blue black.
Red—red lead, vermilion and purple brown, or India red.

Red—crimson, lakes, to which add vermilion or white according to the tone.

Green—grass, verdigrise.

———— invisible, dark ocher, blue, and a little black.

———— a good, patent yellow and Prussian blue.

———— pea, mineral green.

Chocolate India red and black.

Lead Colour—black and white.

Brown—umber raw and burnt.

———— mix black, red, and dark ocher.

Purple—mix lake blue, and white.

Yellow and red lead, make an *orange* colour.

Red and blue make a *purple and violet* colour.

Blue and yellow make a *green* colour.

Black, blue, white, and a little India red make a *pearl* colour.

Light ocher, Prussian blue, and a little black make an *olive* colour.

India red and white, make a *flesh* colour

White and umber, make a *stone* colour.

SMITHING.

SMITHING is the art of uniting several lumps of iron into one mass, and of forming any lump or mass of iron into any intended

———

§ 1. *Description of the Forge.* PL. 33.

The forge consists of a brick hearth raised about two feet six inches, or sometimes two feet nine inches from the floor; heavier work requires a lower forge than lighter work: its breadth must also depend upon the nature of the work; the brick-work may be built hollow below for the purpose of putting things out of the way. The back of the forge is carried up to the top of the roof, and is enclosed over the fire in the form of a funnel to collect and discharge the smoke into the flue, the funnel is very wide at its commencement, but decreases rapidly to the flue, whence it is carried up of a proper size to take off the smoke. The wide part is called the hood or hovel, which in modern forges, particularly in London, is constructed of iron. The air drawn in by the bellows is communicated to the fire by means of a taper pipe, the small end of which passes through the back of the forge, and is fixed into a strong iron plate, called a tue-iron or patent back, in order to preserve the bellows and the back of the forge from the injuries of the fire. A trough for coals and another for water are placed on one side of the forge, generally extending the whole breadth. *See the Plate.*

The best position of the bellows is on a level with the fire-place, but they are frequently placed higher for the purpose of getting room below.

The Tools are as follow.

---

§ 2. THE ANVIL, Pl. 33. Fig. G.

Is formed of a large block or mass of iron with a smooth horizontal face on the top, generally hollowed upon three sides, and on the fourth has a projecting part of a conic figure called a pike, or beckern, or beak iron. The face must be made of steel, so hard as to be incapable of being filed. The anvil is fixed upon a wooden block in order to keep it steady.

---

§ 3. THE TONGS, Pl. 33.

Are of several forms, straight and crooked nosed : the former are used in short flat work, and the crooked nosed in the forging of bars. The chaps, or parts which hold the iron, are placed near the joint, and in order to keep it with greater firmness, a ring is slipped over the ends of the handles of the tongs.

---

§ 4. HAMMERS

Are of several kinds, as hand-hammers, which are of different sizes, according to the weight of the work ; the up-hand sledge is used by under workmen, when the work is not of the largest kind in battering, in order to draw it out to its required dimensions, and for this purpose both hands are used. The about-sledge is the biggest of all the hammers, also used by under workmen in battering the largest work : the former hammer is only lifted up and

T 2

down, but this is slung entirely round with both hands nearly at
the extremity.   The riveting hammer is the smallest of all ; it is
not used at the forge, but in riveting, as its name implies.

---

§ 5. THE VICE, Pl. 34. Fig. B.

Is used to hold any piece of iron or work for the purpose of
bending, riveting, filing, polishing, &c.   It must be placed firmly
and vertically on the side of the work bench, with its chaps parallel to the edge of the said bench.   The inner surface of the chaps
is roughed with teeth, and well tempered ; there is a spring which
acts against the screw pin and opens the chaps ; the screw pin is
cut with a square thread, as also the screw, which is brazed into
the nut box.

---

§ 6. THE HAND-VICE

Is of two kinds, viz. the broad-chapt hand-vice, and the square
nosed hand-vice.   The office of the former is to hold small work
in the act of filing ; it is held in the left hand, and the parts of the
iron turned successively to the file which is used by the right.
The square nosed hand-vice is seldom used, but in filing small
globulous work.

---

§ 7. THE PLYERS

Are of two kinds, flat nosed and round nosed: the former is
used to hold small work while it is fitting to its place, and the
latter for turning or bending wire, or small plates.

§8. DRILLS, Pl. 34. Fig. E.

Are used in boring holes which cannot be punched, owing to the thickness of the iron, or which require more exactness than can be performed by the punch, which is very apt to set the work out of order and shape. Drills are required of various sizes, and to be made of the best steel. The drill consists of a cutting point, a shank, and drill barrel, which must be of a diameter sufficient to turn the drill with the required velocity. The drill is turned by a bow and string, the string is coiled round the barrel, the bow goes with a reciprocating motion, and causes the drill to perform several revolutions in each progressive and retrogressive motion of the bow, and different kinds of work will require different bows, according to the force required to turn the drill, for lighter or stronger work: there is also a drill plate or breast plate, in which the blunt end of the shank of the drill is inserted, and by which the drill is pressed to the work.

To make large holes, more force is required than can be given by the bow and string, instead of which a brace similar to that used by joiners is employed, and the drill itself is fitted in as a bit, instead of the end of the stock, which remains stationary while the other part is turning; there is a long tapering spindle of iron, which is carried round with the brace; the upper end of this spindle is inserted in the lower horizontal side of an iron plate, which is fixed to the under side of a beam, called the drill beam. The drill beam turns upon a transverse pin horizontally posited at one end, and is drawn down. by a weight at the other, and thus presses the brace downwards by the ponderosity of the beam and that of the weight, while the brace is revolved by hand. A piece of iron being laid under the drill bit, where the hole is intended, and the drill turned swiftly round will be bored through, or to any required depth. *See* Plate 34, Fig. E.

§ 9. SCREW PLATES

Are plates of well tempered steel with several cylindric holes of different diameters, with screw threads wrought into square grooves from the surface of the interior concavity; to these plates belong as many pins, tapering to their ends, called taps, which are the frustrums of cones, not differing materially from cylinders : the convex surface is threaded in the same manner, and made to fit their respective holes.

------

§ 10. SHEARS,

An instrument for cutting iron, consisting of two equal and similar pieces moveable round a joint, near to two of the ends, and may be considered as a double lever, so that when two of the ends are opened or shut, the other ends will be opened or shut also. The cutting edges which meet each other are brought to an acute angle, and the surfaces of the inner faces gradually come more and more in contact in the same plane, as the longer ends which are employed as handles are brought nearer together. Shears are used in cutting iron plates, and even bars, and are consequently of various sizes, according to the stiffness or strength of the iron to be cut. When the shears are used, one handle is screwed fast in the vice, and the other only is moveable ; the iron to be cut is laid between the edges which close together.

------

§ 11. SAWS

In general oave been sufficiently defined in § 45 Joinery. They are used by smiths to cut pieces of iron or bars of all dimensions, and for cutting grooves and notches to any required depth. Shears have an advantage over saws in cutting with more rapidity, but saws cut with more exactness, and save the whole or much labor in filing; and may be also used in cutting bars or pieces of the greatest dimensions, where shears cannot be used. Smiths' saws must be very narrow and stiff, with a bow of iron, by which the ends are made fast, and the plate stretched by a screw at one end; the bow has a projecting part in a straight line with the saw, which forms the handle.

§ 12. *Of Forging.*

In forging, the fire must be regulated by the size of the work, and in heating the iron, beat the coals round the outside of the fire close together with the slice, in order to prevent the heat from escaping as often as the flame begins to break out, and in order to save fuel, wet or damp the outside of the coals: to know whether the work takes the heat, draw it a small degree out of the fire, and thrust it quickly in again if not hot enough: if the iron be too cold the hammer will make no impression upon it, or in the language of smiths, it will not batter; if too hot it will break or crack.

§ 13. *Of Heats.*

Heats are of several kinds, depending on the destination of the work, as blood-red heat, white-flame heat, and sparkling or welding heat. The blood-red heat is used when the shape of the iron is not required to be altered, and when the surface is only required to be smooth hammered: this operation is performed by the hand-hammer with light flat blows until the protuberances and hollows are brought to the required surface, whether planed or curved, the work is then prepared for the file. The hammering of the work to a true surface, will save much trouble in filing.

The white-flame heat is used in forming the iron from one shape to another; in the execution of this, one, two, or more men must be employed to batter the work with sledges, until it acquires

2 к

nearly its proposed form and size ; afterwards smooth it with the hand hammer.

A sparkling or welding heat is used when the iron is required to be doubled, or two or more pieces consolidated, in order to make the piece of the required dimensions. In joining two or more bars together, heat them to that degree as to be nearly in a state of fusion ; they must then be taken out of the fire with the utmost despatch, and the scales or dirt which will hinder their incorporation being scraped off, put the pieces in contact at the heated part, and hammer them together until there is no seam or fissure left : this operation will require two or more men, according to the magnitude of the bars. If the particles of the iron have not been sufficiently incorporated by the first heat, more heats and the operations of hammering must be repeated until the work is perfectly sound ; after which it is formed into the shape proposed, and finished by smoothing, &c. To make the iron come sooner to a welding heat, stir the fire with the hearth staff, and throw out the cinders the iron may have run upon, as they will prevent the coals from burning ; to prevent the iron melting, throw some sand over it while in the fire. In this operation care must be taken to prevent the iron from running, which will make it so brittle as to prevent its forging, and so hard as to resist the action of the file. In welding, some smiths strew a little sand upon the face of the anvil, as they conceive it makes the iron incorporate better. If by ill management the iron be wrought too thin or too narrow, and should there be substance enough to make it thicker, give it a flame heat, and set the heated end upright upon the anvil, and hammer upon the cold end until the heated end be beat to the size or turned into the body of the work ; the part so beat is said to up-set, and the operation is called up-setting. When your work is forged, let it cool gradually, and do not by any means quench it in water, which will harden it too much.

§ 14. *To punch a Hole.*

Take a punch of the size and shape of the hole required, the point or narrow end of it must be hardened without tempering, as the heat of the iron will soften it sufficiently, and sometimes too much, and then it must be re-hardened : if the work is not very large, bring the iron to a blood heat, but if very large, bring it almost to a flame heat, and lay it upon the anvil : and place the point of the punch at the spot where the hole is to be made, then with the hammer punch the hole. If the work is very heavy, fix the punch in a wooden rod, and place it on the intended situation of the hole; let another person strike till the punch is forced about half way through, then reverse the iron and punch through on the contrary side; the hole is afterwards smoothed, and per-fected by a mandrill being driven through. But in punching take care to plunge the punch into water as often as it is heated, or as often as it changes colour, in order to re-harden it otherwise it will spoil both the work and the punch.

§ 15. *Filing and Polishing.*

Filing is the operation of cutting or tearing iron in particles or very small parts, called filings, by means of an instrument toothed all over its surface; the instrument itself is called a file. Files are differently formed, and of various sizes for different purposes, their sections being either square, oblong, triangular, or seg-mental ; the files of these sections are respectively denominated square, flat, three square, and half round; they also differ in the magnitude of their teeth, as the iron may be required to be more or less reduced in a given time : it is evident that in the operation of filing, the surface of the iron will be full of scratches, and these scratches will be larger or smaller according as the teeth of the files are coarser or finer : files have therefore obtained the follow-ing names, according to the number of teeth cut on the same area : the largest rough tool file is called a rubber, and is used after the hammer in taking away the prominent parts on the sur-face of the iron ; the bastard tooth file is employed to take out the marks made by the rubber, the fine-toothed file is employed in taking out the scratches made by the bastard-toothed file ; and lastly, the smooth-toothed file is employed in taking out the scratches of the last : the surface is at last made perfectly smooth by means of emery and tripoli. And whatever be the surface of the work, whether flat, cylindrical, or conical, the file must always be made to describe that surface as near as the hand and judgment can direct : these matters, by keeping the principle of motion in view, are soon obtained by practice.

After the surface of the iron has been smoothed by the emery and tripoli, it is then polished by a piece of very hard and highly polished steel, called a burnisher, with a handle at one or both ends, according to the pressure required, which will depend on the magnitude of the surface. The sides of the burnisher are either flat or convex, according to the surface to be polished.

----

§ 16. *To cut thick Iron Plate to any required Figure.*

Having drawn or scratched the figure upon the surface of an iron plate, place it on the anvil, if large; if small, upon the stake : a chisel being in your left hand, with its edge set upon the mark, strike it with the hammer till the substance is nearly cut through, so as to leave a very thin portion of the thickness below it : observe, if the iron were cut through, the face of the anvil being steel, it will batter or break the edge of the chisel, and for this reason when the edge comes very near the under side of the plate, strike only with light blows : repeat this operation till the whole of the figure is gone over ; the part intended to be taken away, may be broken off with the fingers or with a pair of plyers, or by pinching the plate in the vice, with the cut part close to the chaps, and then wriggle it, till it comes asunder.

§ 17. *Riveting*

Is the art of fixing the end of a pin into a hole, by battering or spreading the end which has passed through the hole, so as not only to fill the hole, but to increase its diameter on the opposite side, and thereby prevent its being drawn out again.

---

§ 18. *To rivet a Pin to a Plate or Piece of Iron.*

Having formed the shank to the size of the hole, with a shoulder, and something longer than the thickness of the plate, file the end of the shank flat, so that it may batter more easily ; slip the shank into the hole, and keeping the shoulder in contact with the surface of the plate; the end of the pin abutting upon the stake, and the pin standing perpendicular, strike the edge of the end of the shank with light blows, until it is spread all round, then lay heavier blows, sometimes with the face, and sometimes with the pen of the hammer, till the end of the shank is sufficiently battered over the plate : in performing this operation, care must be taken to keep the pin at right angles to the plate, and the shoulder close.

---

§ 19. *To make small Screw-Bolts and Nuts.*

Supposing the shank of the screw-bolt to be let into a square hole, in order to keep it from twisting by the turning of the nut; take a square bar or rod of iron near the size of the head of the screw-pin, and bring it to a flame-heat; take as much of the length of the bar as is equal to the length of the shank, and lay one side flat upon the nearer side of the anvil, and hammer it down to the intended thickness; this will forge two of the sides at once, the under side being forged by the anvil, and the upper beat flat with the hammer; but if the iron get cold before the forging is finished, it must have another heat. Then lay one of the un-wrought sides upon the nearer side of the anvil, and hammer this side straight as before, so that the two other sides will also be made; then beat in the angles so as to make it nearly round, and of such length as is equal to the intended length of the screw pin. Having forged the shank square, and formed the head either square or round as may be intended; file also the screw pin so as to make it taper in a small degree, and to take out the irregu-larities of the forge; the conic form makes it enter more easily, and the irregularities being taken away, makes the screw more exact in the distances of the threads : the quantity of taper may be something more than twice the depth of the threads. Then fix the bolt with the head downwards into the vice, and with a screw plate equal to the interior diameter of the cylinder from which the screw is to project, lay the hole upon the end of the screw pin, and press it hard downwards. Then turn the screw plate parallel to the horizon from right to left with a uniform pressure round about the pin, both progressively and retrogressively, and the plate will begin to groove out the channel between the thread of the screw : proceed with this process until as much of the screw be formed as is required.

To make the nut, the hole must be equal to the diameter of the cylinder from which the thread is made in the shank of the screw, and the tap must be made tapering, in order to enter the hole. Proceed and screw the nut in the vice, with the axis of the cylin-dric hole vertical, and enter the screw tap, which turn by the handle as before, and it will begin to cut the interior groove of the nut; proceed working until the groove between the thread be of its full depth : the thread and groove in the nut will thus be made to fit the groove and thread of the screw pin.

§ 20. *Of Iron.*

Iron is a metal of a bluish white colour, of considerable hardness, but easily formed into any shape, and is susceptible of a very fine polish. It is the most elastic of all the metals, and next to platina, is the most difficult of fusion. Its hardness in some states is superior to that of any other metal, and it has the additional advantage of suffering this hardness to be increased or diminished at pleasure, by certain chymical processes, without altering its form. Its tenacity is also greater than that of any other metal, except gold; an iron wire, the tenth part of an inch in diameter, has been found capable of sustaining more than 500lb. weight without breaking. Its ductility is such as to allow it to be drawn into wire as fine as a hair.

Iron ore is found mixed with sand, clay, chalk, and in many kinds of stones and earths. It is also found in the ashes of vegetables, and the blood of animals in great abundance. Iron ores are therefore extremely numerous.

Iron is obtained from the ore by an operation called smelting, and in this state it is called crude iron, cast iron, or pig iron, but it is very impure. Cast iron is scarcely malleable at any temperature, it is generally so hard as to resist the file, and is extremely brittle; however, it is equally permanent in many applications with wrought iron, and is less liable to rust; and being easily cast into various forms by melting, is much cheaper. Indeed the labour to wrought iron if applied to many of the purposes to which cast iron is used would be incredible, and in some cases insurmountable. The use of cast iron is sufficiently obvious in the wheel work of every department of machinery, in crane work, in iron bridges, in beams and pillars for large buildings, and in numerous articles of manufacture.

Cast iron is reduced into wrought or bar iron, or forged iron, by divesting it of several foreign mixtures with which it is incorporated. The varieties of wrought iron are the following: hot-short iron is so brittle when heated that it will not bear the weight of a small hammer without breaking to atoms, but is malleable when cold, and very fusible in a high temperature; cold-short iron possesses the opposite qualities, and is with difficulty fusible in a strong heat, and though capable while hot of being beaten into any shape, is when cold very brittle, and but slightly tenacious. The iron in general use, which though in a chymical point of view is not entirely pure, is so far perfect that it possesses none of these defects; its principal properties are the following: 1st. When applied to the tongue it has a styptic taste, and emits a peculiar smell when rubbed: 2d. Its specific gravity varies from 7·6 to 7·8; a cubic foot of it weighs about 580lb. avoirdupoise: 3d. It is attracted by the magnet or loadstone, and is itself one of its ores, the substance which constitutes the loadstone. It is also capable of acquiring itself the attraction and polarity of the mag-net in various ways; iron, however, that is perfectly pure, retains the magnetic virtue only a very short time: 4th. It is malleable in every temperature, which as it rises, increases the malleability. It cannot, however, be hammered out so thin as gold or silver, or even copper. Its ductility is very great, and its tenacity is such, than an iron wire something less than the twelfth of an inch in diameter, is capable of supporting without breaking 549¼lb. avoir-dupoise: 5th. it melts at about 158° of Wedgewood: 6th. it com-bines very readily with oxygen; when exposed to the air its surface is soon tarnished, and is gradually changed into a brown or yellow colour, usually called rust: this change takes place more rapidly, as it is more exposed to moisture.

To preserve iron from rust, particularly when polished, various methods have been tried with more or less success: among others, the partial oxidation, known by the term bluing, has been adopted; the slightest coat of grease is sufficient to prevent rust.

Iron is the most useful and the most plentiful of all metals. It requires a very intense heat to fuse it, on which account it can only be brought into shape of tools and utensils by hammering: this high degree of infusibility would prevent the uniting of several

masses into one, were it not from its being capable of welding, a
property which is found in no other metal except platina. In
a white heat, iron appears as if covered with a kind of varnish, and
in this state, if two pieces be applied together, they will adhere,
and may be perfecfly united by forging.

Steel is made of the purest malleable iron by an operation called
cementation, by which it acquires a small addition to its weight,
amounting to about the hundred and fiftieth or two hundredth
part. In this state it is much more brittle and fusible than before.
It may be welded like bar iron, if it has not been fused or over
cemented ; but its most useful and advantageous property is, that
of becoming extremely hard when heated and plunged into cold
water ; the hardness which it thus acquires is greater, as the steel
is hotter and the water colder. The sign which directs the mechanic in the tempering of steel, is the variation of colour which
appears on its surface. If the steel be slowly heated the colours
which it exhibits are a yellowish white, yellow, gold colour, purple, violet, deep blue. If the steel is too hard, it will not be
proper for tools which are intended to have a fine edge, as it will
be so brittle that the edge will soon become notched : and if it is
too soft, the edge will soon turn aside, even by very slight usage.
Some artists heat their tools and plunge them into cold water,
after which they brighten the surface of the steel upon a stone ;
the steel being then laid upon hot charcoal, or upon the surface
of melted lead, or placed on a bar or piece of hot iron, gradually
acquires the desired colour, and at this instant it must be plunged
into water. If a hard temper is required, as soon as a yellow
tinge appears, the piece is dipped again and stirred about in the
cold water. In tempering of tools for working upon metals, it
will be proper to bring it to a purple tinge before the dipping.
Springs are tempered by bringing the surface to a blue tinge.
This temperature is also desirable for tools employed in cutting
soft substances, such as cork, leather, and the like ; but if the
steel be plunged into water when its surface has acquired a deep
blue, its hardness will scarcely exceed the temperature of iron.

No. 18   2 L

When soft steel is heated to any one of these, and then plunged into water, it does not acquire so great a degree of hardness as if previously made quite hard. The degree of heat required to harden steel, is different in the different kinds. The best kinds require only a low red heat ; the harder the steel, the more coarse and granulated its fracture will be. Steel, when hardened, has less specific gravity than when soft ; the texture of steel is rendered more uniform by fusing it before it is made into bars, and in this state it is called cast steel, which is wrought with more difficulty than common steel, because it is more fusible, and will disperse under the hammer if heated to a white heat. Every species of iron is convertible into steel by cementation ; but the best steel can only be made from iron of the best quality which possesses stiffness and hardness as well as malleability. Swedish iron has been long remarked as the best for this purpose.

The *Cast Steel* of England is made as follows : a crucible about ten inches high, and seven inches in diameter, is filled with ends and fragments of the crude steel of the manufactories, and the filings and fragments of steel works ; they add a flux, the component parts of which are usually concealed. It is probable, however, that the success does not much depend upon the flux. This crucible is placed in a wind furnace, like that of the founders, but smaller, because intended to contain but one pot only. It is likewise surmounted by a cover and chimney, to increase the draught of air ; the furnace is entirely filled with coke, or charred pit-coal. Five hours are required for the perfect fusion of the steel. It is then poured into long, square, or octagonal moulds, each composed of two pieces of cast iron fitted together. The ingots when taken out of the mould, have the appearance of cast iron. It is then forged in the same manner as other steel, but with less heat and more precaution. Cast steel is almost twice as dear as other good steel ; it is excellent for razors, knives, joiners' chisels, and for all kinds of small work that require an exquisite polish : its texture is more uniform than common steel, which is an invaluable advantage. It is daily more and more used in Eng-

land, but it cannot be employed in works of great magnitude, on account of the facility with which it is degraded in the fire, and the difficulty of welding it.

To conclude: British cast iron is excellent for all kinds of castings; our wrought iron also of late has been much improved in the manufacture, and by many persons is thought not to be inferior to that of Sweden, which till lately had a decided preference, and is to be attributed to the use of charcoal in the process of smelting, which cannot be procured in sufficient quantity in England, where pit coal has of necessity been substituted. The Navy Board and East India Company, however, now contract for British iron only.

PLATE XXXIII.

Perspective View of a Smith's Work Shop, shewing a double Forge with its Apparatus, and some Tools in general Use.

A back of the forge.

B the hood.

C Bradley's patent back, showing the nozel or the iron of the bellows.

D end of the forge.

E bellows with the rock staff.

F troughs for coals and water.

G anvil, shewing the beak iron, and a hole for holding the tools on the top. The anvil being supported upon a wooden block.

H a strong stool for supporting the chasing tool I.

I the chasing tool for rounding bolts, and punching holes in iron; the holes are called bolsters, and those upon the sides are called rounding tools; the whole is called generally a bolster.

K a sledge hammer.

Near D is a horse to hold up long pieces of iron at the end of the forge, when found necessary.

The square hole near A is used for discharging the ashes, which slide down a hollow, and come out at the bottom of the front.

The coal trough is placed next to the forge, and the water trough next to the front. The tongs are shewn in the water trough, and a pair of lip and straight tongs are shewn on it.

In smiths' shops, where heavy articles are manufactured, cranes are employed for taking the work out of the fire.

PLATE XXXIV.

View of another Part of a Smith's Work Shop, shewing the Work Benches with the Vices, the Drill in the act of Boring, and a Turning Machine, as wrought by a Winch and Wheel, as also by the Foot.

A, A work benches.

B, B, B vices.

C the bench anvil.

D, E, F, G various parts of a drill machine.

D the drill block.

E the drill and brace.

F the drill beam, shewing the lever to pull it up.

G a rod to hang a larger or smaller weight, for giving more or less power to the drill, as may be required in boring a greater or less hole.

H, I, K, L parts of the turning lathe.

H handle to turn the large wheel.

I the large wheel.

Pulleys for the cord.

L puppets, rest, collar, and mandril.

N wheel and crank for revolving the mandril by the foot, &c

W.S. Barnard. Sc.

INDEX

EXPLANATION OF TERMS

USED IN

SMITHING.

N. B. *This Mark § refers to the preceding Sections, according to the Number.*

------◆--

A.

ABOUT SLEDGE, the largest hammer used by smiths; it is slung round near the extremity of the handle, generally used by under workmen, § 4.

ANVIL, a large block or mass of iron with a very hard smooth horizontal surface on the top, and a hole at one end of the surface, for the purpose of inserting various tools, and a strong steel chisel, on which a piece of iron may be laid and cut into two. Anvils are sometimes made of cast iron, but the best are those which are forged, with the upper face made of steel. Small anvils are also used in more delicate parts of the business, § 2. *See Plate* 33. *Fig.* G. *Plate* 34. *Fig.* C.

B.

BAR IRON, long prismatic pieces of iron, being rectangular paral-lopipeds, prepared from pig iron, so as to be malleable for the use of blacksmiths. For the method of joining bars, see § 13

BASTARD CUT, § 15.

BASTARD-TOOTHED FILE, that employed after the rubber, § 15.

BATTER, to displace a portion of the iron of any bar or other piece by the blow of a hammer so as to flatten or compress it inwardly, and spread it outwardly on all sides around the place of impact.

BEAK IRON, the conic part of the anvil, with its base attached to the side, and its axis horizontal, § 2. *See Plate* 33. *Fig.* G.

BELLOWS, the instrument for blowing the fire, with an internal cavity, so contrived as to be of greater or less capacity by reciprocating motion, and to draw in air at one place while the capacity is open upon the increase, and discharge it by another while upon the decrease. The bellows are placed behind the forge, with a pipe of communication through the back to the fire, and are worked by means of a lever, called a rocker. *See Plate* 33. *Fig.* E.

BENCH, an immoveable table, to which one or more vices are fixed, for filing, drilling, and putting work together. *See Plate* 34.

BLOOD-RED HEAT, the degree of heat which is only necessary to reduce the protuberances of the iron by the hammer, in order to prepare it for the file, the iron being previously brought to its shape. This heat is also used in punching small pieces of iron, § 13.

BOLSTER, a tool used for punching holes, and for making bolts. *See Plate* 33. *Fig.* 1.

BRACE, an instrument into which a rimer is fixed, also part of the press drill.

BREAST PLATE, that in which the end of the drill opposite the boring end is inserted, § 8.

BRITTLENESS in iron is a want of tenacity or strength, so as to be easily broken by pressure or impact. When iron is made too hot, so as to be nearly in a state of fusion, it becomes so brittle as to prevent forging, and so hard as to resist the action of the file. This is also the disposition of cast iron.

BROAD CHAPT HAND-VICE, § 6.

Burnisher, an instrument used in polishing, § 15.

C.

Callipers, a species of compasses, with legs of a circular form, used to take the thickness or diameter of work, either circular or flat ; used also to take the interior size of holes.

Cast Iron, § 20.

Cast Steel, § 20.

Cementation, is the process of converting iron into steel, which is done by stratifying bars of iron in charcoal, igniting it, and letting it continue in a kiln in that state for five or six days, by which the carbon of the charcoal is absorbed by the iron, and causes it to become steel.

Chaps, the two planes or flat parts of a vice or pair of tongs or plyers, for holding any thing fast, which are generally roughed with teeth.

Chisel, a tool with the lower part in the form of a wedge, for cutting iron plate or bar, and with the upper part flat, to receive the blows of a hammer, in order to force the cutting edge through the substance of the iron. For its use, see § 15.

Cold Short Iron, iron in an impure state, § 20.

Compasses, an instrument with two long legs, working on a centre pin at one extremity ; used for drawing circles, measuring distances, setting out work, &c.

Counter-sink, a tool used to make the necessary bevel, to admit the head of a screw, rivet, &c. *See Joinery*, § 36.

Crooked Nosed Tongs, § 3.

D.

Draw, to draw is the act of lengthening a bar of iron by hammering, also wire reduced from any size to a smaller, is said to be drawn.

Drill, a boring tool which forms a cylindric hole with the greatest exactness. Drills are particularly used where the substance is too great for the operation of the punch, or where very exact cylindric holes are required, § 8.

DRILL Bow, § 8.

E.

EMERY, a very fine powder, prepared from iron, used in polishing, § 15.

F.

FILE, § 15.

FILING, § 15.

FINE-TOOTHED FILE, § 15.

FLAME HEAT, is that which is required in forming the iron from its original shape. This degree of heat is also required in upsetting, § 13.

FLUX, any substance, which, mingled with a body, accelerates its melting. Fluxes are salt, bone-ash, charcoal, lime-stone, borax, &c.

FORGE, to form a piece of iron into any required figure or shape, by means of heat and the hammer, or to weld several pieces of iron, § 13.

FORGE, the furnace for heating the iron so as to become malleable, and thence prepare it for forging, § 1.

G.

GAUGE, an instrument for taking the size of any bar, &c. made from one-eighth of an inch to any size, is a piece of iron with regular notches of the sizes required.

GRIND-STONE, used for sharpening tools, &c. used also previous to the file in many cases.

H.

HAMMERS used by smiths are of four kinds, viz. the hand-hammer, the up-hand sledge, the about sledge, and the riveting hammer, § 4.

HAND HAMMER, that which is held by one hand while the iron is held by the other, for smoothing work. Hand hammers are of different sizes, § 4.

HAND VICE, used for turning about small pieces of iron, while filing on the large vice, which would otherwise be too small for the hand to command with sufficient power, § 6

HEARTH STAFF, a bar or poker of iron for stirring the fire.

HEATS, the several degrees or intensities of heat necessary for performing certain operations of forging. Heats are of three kinds, viz. blood-red heat, white-flame heat, sparkling or welding heat, § 13.

HOOD, the lower part of the chimney, expanding in its horizontal dimension downwards from the flue to its mouth, which is considerably above the hearth of the forge. *See Plate* 33. *Fig.* B.

HOT SHORT IRON, iron in an impure state, § 20.

HOVEL, the same as Hood.

I.

INGOT, a mass of metal.

IRON, the material used by smiths, § 20. Ornamental work, such as brackets and lamp irons, is charged at least one third more than plain hammered work, such as rails, window bars, &c. and sometimes more than twice the sum, according to the quantity of ornament.

L.

LATHE, an instrument used in turning rounds, ovals, &c. *See Plate* 34. *Fig.* H.

M.

MANDRIL, a cylindric pin of iron, used to perfect a hole after the

2 M

punch; also a conical tool of iron three or four feet high, used for making rings, or other circular work; also a part of the turning lathe.

N.

Nippers, an instrument like a pair of pinchers, with sharp edges, used to cut iron wire, &c.

Nut of a Screw, a piece of iron pierced with a cylindric hole, the circumference of which contains a spiral groove. The internal spiral of the nut is adapted to an external cylindric spiral on the end of a bolt. The use of the bolt and nut is to screw two bodies together, a head being wrought on one end of the bolt, in order to counteract the action of the nut. By this means the two bodies are held together by compression, and the bolt between the head and the nut becomes a tie, § 19.

P.

Pig Iron, short thick bars of iron, in the state in which it comes from the smelting furnace.

Plate or Sheet Iron, plates of iron flattened by a roller, of various sizes and thickness. m

Plyers, small tongs for holding small pieces of iron, § 7.

Punch, a kind of chisel with two flat ends for piercing iron by a hammer, one end which has the greater area receives the blows of the hammer, and the other, which has the less, makes its way through the iron, and forms a hole, § 14.

R.

Red Sear, is when the iron is made so hot as to crack by the hammer.

Rimer, a tapering instrument, square, triangular, &c. used to enlarge holes. *See Joinery*, § 37.

Rivet, to fasten the end of a pin or bolt by battering the end of it.

Rock Stiff, or Rocker, the lever which gives motion to the bellows.

Rod Iron, small bars of iron, square, round, or flat.

Rounding Tool, a tool used for rounding a bar of iron, of two pieces, each with a semi-circular cavity, according to the size wanted; one piece is fixed into the anvil, while the other, held by a rod or handle, is applied over the iron, and is struck with a hammer.

Rubber, the file which is first used upon the iron in reducing the protuberant parts left by the hammer; it has fewer teeth on the same area than any other file, § 15.

———

S.

Saws, § 11.

Scales, the laminated parts accumulated on the surface of the iron by heat.

Screw, a pin with a spiral groove cut within the surface of a cylinder, and with a nut having a hole adapted thereto, § 19.

Screw Driver, a tool used to turn screws into their places.

Screw Plate, that which cuts the spiral groove within the cylindric surface of the pin, § 9.

Screw Threads, the parts which are left standing between the spiral grooves of the screw.

Shears, § 10.

Shut, the same as weld, which see.

Side Set, a hammer used to set shoulders of rivets to a true square or bevel, as required.

Slice, the instrument for beating the fire close.

Smooth-toothed File, the finest of all the files. and the last used in polishing the surface, § 15.

Sparkling Heat, the intensity necessary in welding two or more pieces of iron together, § 13.

Square, an instrument used to examine if the work be done to a right angle; for a particular description, *See Joinery*, § 36. The smith's square is all iron.

SQUARE-NOSED HAND VICE, § 6.

STEEL, § 20.

SWAGES, all instruments used to give the form or contour of any moulding, &c. used in the same manner as the rounding tool.

T.

TAP, a tapering pin of the form of a conic frustum, approaching very nearly to a cylinder, with a spiral groove cut on its surface, for making the interior or female spirals of a screw nut, § 9.

TAP-WRENCH, an instrument used to turn the tap in making screws.

TONGS, an instrument with long handles, used for holding pieces of hot iron in the operation of forging. Some are straight nosed, others crooked nosed.

TRIPOLI, a species of argillaceous earth, reduced to a very fine powder, and used in polishing the finest works, is also used in polishing marbles, minerals, &c.

TUE IRON, the plate on the back of the forge, which receives the small end of the taper pipe, which comes from the bellows for conveying the stream of air to the fire.

U.

UP-HAND SLEDGE, § 4.

UP-SETTING, § 13.

V.

VICE, an instrument for holding any thing fast, § 5.

W.

WASHER, the instrument for damping the fire.

Washer, a piece of flat iron with a hole placed between the nut of a screw and the wood, to prevent the wood being gulled.

Welding, is that intimate union produced between the surfaces of two pieces of malleable metal when heated almost to fusion and hammered. This union is so strong, that when two bars of metal are properly welded, the parts thus joined are relatively as strong as any other part. Only two of the old metals were capable of a firm union by welding, namely, platina and iron, the same property belongs to the newly discovered metals, potassium and sodium.

Welding Heat, the same as sparkling heat, § 13.

White-flame Heat, the intensity necessary in forming a piece of iron into another shape, § 13.

Wrench, a forked instrument used in screwing up of nuts.

TURNING.

§ 1. TURNING in general is the art of reducing any material to a certain required form, by revolving the material according to a given law, in a machine called a lathe, and cutting away the superfluous substance with a gouge or chisel, which is held steady upon a rest, until the surface be sufficiently reduced : sometimes pressing the cutting edge gently forwards, and sometimes sidewise, according to the design, until it has obtained the figure and dimensions required.

The art of turning is of very remote date. The invention is ascribed by Diodorus Siculus to Talus, a grandson of Dædalus ; but Pliny says it was invented by Theodore of Samos, and mentions one Thericles as being famed for his dexterity in this art. By means of the lathe, the ancients formed vases, which they enriched with figures and ornaments in basso relievo.

The Greek and Latin authors make frequent mention of the lathe ; and it was a proverb among them to say a thing was formed by it when the parts were delicate, and their proportions correct.

Turning is performed either by the body being continually revolved, or by the rotation being made backwards and forwards ; but the latter mode is attended with a loss of time.

The materials employed in turning, are wood, ivory, brass, iron, stone, &c.

Turning is also of different kinds, as *circular turning, elliptic turning, and swash turning ;* these may be said to be the simple movements of the machine, according to geometrical principles, but by means of moulds, an indefinite number of things may be formed in this way ; but in all of them, suppose for a single revolution of the machine, the cutting edge of the instrument is held immoveable to the same point of space, and the machine is so regulated, as to bring the different parts of the intended surface to the cutting edge in its revolution. In practice, instead of the cutting edge of the instrument being exactly at the same place when a considerable surface is to be wrought, it is made to traverse the surface, that is, to have a slow lateral movement in the direction of the intended form, and by this means to shave off spiral turnings.

———

§ 2. *Circular Turning*

Is the art of forming bodies of wood, ivory, metal, stone, &c. by revolving the body upon a given straight line, as an axis in a machine, while the cutting edge of a tool is held at such distance, as to cut or shave off the prominent parts in thin slices, as the body revolves, until it acquires the intended form.

From the definition here given, it is evident, that all points of the solid in the act of turning, will describe the circumference of circles in planes, perpendicular to the axis, which will pass through their centres.

Every section passing through the axis of the turned body, will have the two parts on each side of the axis equal and similar figures: and any straight line perpendicular to the axis, and terminated by the sides of the section, would be bisected by the said axis.

For the sake of perspicuity, we shall call any section through the axis, the axal section, that is, a section of the body in which the axis would be entirely in its plane ; the design of the turning depends entirely upon this section, which, if it be a circle, the body when turned will be a sphere, and if an ellipse, it will be a spheroid, &c. This is the most useful of all kinds of turning, and essential in the construction of many kinds of engines and machinery, where every other method would fail, as not being sufficient to give the desired accuracy. Its use in fancy work is beyond description, and the labour thereby rendered easy. The practice will be obtained better from actual practice of the business, than from any description.

The following are the descriptions of the most useful wood lathes, which have the same principles in common with those for turning metals.

---

§ 3. *Lathes in general.*

Lathes are of several kinds, as the *pole lathe*, the *foot lathe*, and the *wheel lathe*, which is used in very large work, and is revolved by manual strength. It consists of a great wheel with a winch handle at the end of its axle, by which the force is communicated. There are other lathes used for very large work, driven either by steam engines, water wheels, or by horse power. All these ought to be so contrived, that the works may be stopped, even though the power be still exerted.

---

§ 4. *The Pole Lathe.*

The pole lathe consists of the following parts, several of which are common to every other description; the legs or stiles for supporting it, the shears horizontally fixed with a parallel cavity between them for conducting the puppets; the puppets sit vertically, and are made to slide between the cheeks of the shears, the one being made to receive the screw, and the other to receive the conical point, which is fixed horizontally in one puppet for supporting one end of the piece to be turned in its axis, the screw with another point supporting the other end of the piece to be turned, by means of the screw, the body may be fastened or slack.

ened at pleasure ; the rest for the tool fixed horizontally to the puppets, and parallel to the cheeks, the tenons made on the lower end of the puppets, in order to form a shoulder for re-acting against the wedges below, the wedges for fastening the puppets so as to regulate them to any distance ; the treadle and cross treadle for the foot, in order to give a reciprocal rotation to the body to be turned, by means of a string coiled round it, and an elastic pole which re-acts against the string and the pressure of the foot ; the pole for pulling up the treadle and acting reciprocally against the pressure of the foot, the string for turning round the body by the pressure of the foot downwards, and the re-action of the pole upwards.

The legs or stiles may be about two feet ten inches high, and are tenoned into the cheeks at their upper ends, and fixed by pins or screws, the latter is preferable. In turning large work, it will be necessary to brace the legs and cheeks to the floor or ceiling, as may be found convenient, otherwise the work will be liable to tremble. The puppets are pieces of a square section, and ought to be sufficiently strong to answer every description of work.

The pole lathe is used in turning heavy or long work, the string is coiled round the material, which performs the office of a mandrel : but for general use this kind of lathe is not so convenient as that which is called the foot lathe ; and besides this, there is a loss of time in making the alternate revolutions. The pole lathe is now but little used. It is sometimes, as well as other lathes, tightened with a screw and washer.

This lathe has two puppets with a pin or centre in each, the right centre is moveable by a screw, but the left puppet with the centre is generally stationary, and the work is supported upon the centres. The rest is moveable between the shears, and fastened by means of a screw bolt. In beginning to operate with this machine, there must be a small part turned, in order to act as a pulley.

No. 19   2 N

§ 5. *Foot Lathe.*

The foot lathe consists of machinery and a frame for sustaining it. The parts of the machinery are the treadle, the crank hook, the great wheel or fly, the band, and the mandrel; the parts of the frame are the feet, the legs, the back board or bench, the pillars, the puppet bar or bed, the puppets, and the rest.

The treadle or foot board is put into alternate motion by the pressure of the foot downwards, and the momentum of the fly-wheel upwards; the board or frame of the treadle is screwed to an axle, on which it turns.

The connecting rod or crank hook is hooked into a staple in the middle of the treadle board, and may be lengthened or shortened at pleasure by screwed hooks; it may either be constructed of iron or brass, but is most frequently of iron, and even sometimes of leather.

The foot wheel or fly is put into motion by means of the treadle and a crank on the arber of the wheel; the motion is communicated from the treadle by the crank hook or connecting rod, and fastened to the crank of the wheel by a collar, embracing and turning round at the upper end. The foot pushes down the treadle, and gives the wheel a rotative motion, and when the crank has been drawn to the lowest point, the momentum which the wheel has thus acquired draws up the treadle, and thus by the alternate pressure of the foot, and the momentum of the wheel, the motion is continued. The wheel was formerly constructed of wood, but now generally of cast iron; the general surface of the exterior side of the rim, is sometimes conical, and cut with three or four angular grooves, which are best when recessed with an angle, so as not to have a flat bottom: this form is advantageous, on account of the band having more power to turn the wheel. Some wheels have two or more rims, in order to give different degrees of velocity or to increase the power. The axle of the wheel is made of wrought iron, except the centres, and bent in the middle, to form

the crank: the centres at the ends are made of hard steel, welded
to the iron part of the axle. The band connects the fly and mandrel, and is mostly made of cat-gut of such thickness as the nature of the work may require. It is either spliced at the joining,
or the two ends fastened together by hooks and eyes; the band
may be either tightened by grooves in the great wheel, or in the
pulley of the mandrel, or by sliding pieces in the legs.

The mandrel consists of an axle and pulley. The axle is constructed of wrought iron, except the part which turns in the collar,
and which ought to be of hardened steel, welded round the iron
part. The whole of the axle of the mandrel ought to be turned
true in a lathe. It receives a supply of oil from a small hole drilled
down from the top of the puppet, and through the steel collar.

The manner of holding the work, is very different and various,
almost in every instance. In general it is held in pieces of wood
called chucks, which are screwed or cemented upon the nose of
the mandrel. The socket for the mandrel to work in has been
generally made in the back screw, but some experienced workmen prefer it to be in the mandrel. The mandrel is sustained at
one end by the back centre, and at the other end by the steel collar in the middle of the puppet head: the right hand extremity,
called the nose, projects over the puppet, and terminates in a
screw, which is sometimes convex, sometimes concave, and sometimes both: but if there is only one, the convex or male screw is
generally preferred. The pulley has generally three or sometimes
four grooves of different sizes to receive the band, and by this
means it may be turned with different degrees of velocity, and
made to accommodate the length of the band. The edge of the
pulley is bevelled in the same degree as the edge of the fly-whee'
and with the same number of grooves, but the lesser diameter of
the pulley is upon the same side as the greater diameter of the
fly-wheel, and consequently, the greater diameter of the pulley
upon the same size as the lesser diameter of the fly-wheel.

The parts of the frame are as follows: the two feet are screwed to the floor, and mortised to receive the legs, which are fixed

thereon.   Sometimes there is only one leg to each foot, but in the best constructed lathes there are two; the top of the legs are tenoned, which are received by the mortises in the bearers at the top, and fixed therein.

The back board is fixed to the bearers, and supports two pillars which are fixed to it, one being at each end in a vertical plane with each leg or pair of legs.   The puppet bar, or bed, or bearer, is fastened at each end into each pillar, with mortise and tenon ; the common foot lathes have no back board, and the bed consists of two parallel parts, called by some shears, the vertical sides of which form a cavity between them.   The puppets are so constructed as to be moveable upon, and fastened to the bar at pleasure, by means of a screw below the bed; they are generally three in number, the two extreme ones of which have pins with centres, and the middle one has a collar for receiving the ends of the mandrel.   In turning of light-work, not very long, the right hand and middle puppets are used, and the work is sustained by a chuck fastened to the end or nose of the mandrel.   In the common lathes, the puppets are made of wood, and tenoned below, to fit the hollow between the shears or bed, and the tenons are made sufficiently long to come below, so as to receive wedges through a mortise cut therein, and by this means to fix them.   In the best constructed lathes, the puppets are made of cast iron, and moveable also upon a cast iron bearer, and fixed to the required distance by a vertical screw underneath, which comes in contact with a horizontal plate or washer below the said bar.   The puppet which receives the end of the mandrel for holding the work has a cylindric hole with a conic shoulder, through its upper end, and with the axis is directed to the centres in the other puppet.   The fore puppet has a cylindric hole through its top, to receive a polished pointed rod, which is moved by a screw working in a collar.   The puppets are made so as to take off the bar at pleasure ; they are made forked below, and saddled upon the two upper sides of the bar.   The sides or prongs are made very stout, and mortised to receive a short iron bar, which encloses the lower part.   Through receive a short iron bar, which encloses the lower part. Through the middle of this bar, a screw passes underneath, and comes in contact with a thin washer or plate on the under side of the bed, to prevent bruising it. In order to move the puppets freely, and to support them firmly, the bed ought to be made very straight, and of sufficient strength to preserve its figure.

The rest is made so as to be moveable round the work, and fixed in any position, and may be conducted and fastened to any part of the bed.

The framing and the machinery are thus connected: the treadle is fixed into the feet, or in brackets fixed in the back angles formed by the legs and the feet: the fly is sustained at each end by a transverse piece moveable up and down in a frame, and made stationary in any part it is moved to, and thus it may either accommodate the length of the band or the crank hook. The mandrel is sustained at one end by the back centre, which is fastened into the head of the left puppet, and the other into the steel collar as before mentioned.

The machinery is thus put in motion : Suppose the crank to be raised about half a revolution from the bottom, then with considerable force pressing the treadle downwards, the fly-wheel will be put in motion, but if the force communicated is not sufficient to carry it round, it must be pressed down in the act of descending, as often as may be sufficient to put it in rotation, in the required direction of motion ; at every time the treadle begins to descend, press with the foot. The momentum which the fly has thus acquired, will be sufficient to carry it round, even though retarded in a certain degree by an obstacle, until it receive an additional impulse by the foot acting upon the treadle ; then by this momentum, and the continued impulses, the motion is continued, even though the force of the tool is continually acting upon the body in the act of working, and therefore continually destroying a part of the force exerted upon the machine ; but the part thus destroyed is always renewed by an equivalent.   The motion being continued, the band communicates the rotation to the mandrel, and the mandrel to the body, which is fastened to the end of the spindle in the manner before described.

---

§ 6. *A Chuck.*

Is a piece of wood or metal made to fasten on the end of the mandrel, and to sustain the material while it is being turned. Chucks are variously constructed, according to the design of the thing required to be turned. They are sometimes made of wood, and sometimes of metal, particularly of brass. Wooden chucks have a cylindric hole, in which the end of the work to be turned is inserted, and are hooped, in order to prevent splitting when the work is driven into the cavity : this kind of chuck is that which is most frequently used. The work is also sometimes cemented to the chuck, and sometimes screwed to it, as the figure of the thing to be turned may require. The end of the chuck which is screwed upon the nose of the mandrel, is sometimes a concave and sometimes a convex cylinder, the superfices being concentric, or having the same axis. In turning small work, such as snuff-boxes, the material is fastened upon a hollow chuck. It is probable, that the name chuck has originated from the work being driven, jammed, or chocked into it.

---

§ 7. *Of Tools.*

The principal tools employed in turning, are gouges, chisels, right-side tools, left-side tools, round tools, point tools, drills, inside tools, screw tools, flat tools, square tools, triangular tools, turning gravers, parting tools, calipers, &c.

§ 8. *The Gouge* (PL. XL. FIG. 1.)

Is used for roughing wood into its intended form ; also in finishing hollows : the cutting edge is rounded. In turning, the gouge
must be held with an inclination, and the handle considerably depressed, so that the side or basil of the gouge comes very nearly in
a tangent to the circumference of the work, or in the tangent of a
less circle, and consequently the cutting edge of the gouge will be
above the axis. In the use of this tool, the rest is generally upon
a level with the axis. Gouges are of various sizes, according to
the work.

§ 9. *The Chisel* (PL. XL. FIG. 2.)

Is used after the work is roughed into form by the gouge to finish
cylindric, conic, or convex bodies. In the use of this tool, the
bank or horizontal part of the rest, is raised considerably above
the centre of the work, so as to be nearly upon a level with the
surface, and the cutting edge must stand obliquely to the axis of
the cylinder, so as to prevent either angle from running into the
work ; the chisel ought to traverse the work gradually, but not too
fast, as otherwise it will leave a roughness on the surface. This
tool is used principally for soft wood. The basil must be made
from both sides. Chisels are of various sizes, from a quarter of
an inch to two incnes and a half: these are convenient in running
mouldings and cleaning the bottoms of grooves.

§ 10. *Right-Side Tools* (PL. XL. FIG. 3.)

Are used for turning of cavities of hollow cylinders, or those
hollows which have only one internal angle in turning both the

bottom and the side : for this purpose, the tool is made to cut both by its end and side-edge, so that these two cutting edges form an angle with each other rather acute. This tool must be held on a level with the axis of the work. Side tools are made of different widths, to suit various cavities. The basil is only made from one side of the tool. The flat side is upwards, and consequently, the basil downwards

----

§ 11. *Left-Side Tools*

Are not used in internal work, as the right-side tools, but up the left side of convex surfaces, such as spheres, torus mouldings, ovolos, &c. The acute angle is upon the contrary side of this tool to the other. Left-side tools are likewise made to various widths.

----

§ 12. *Round Tools* (PL. XL. FIG. 4.)

Are used for turning concave mouldings, and are of various widths, to adapt themselves thereto.

----

§ 13. *Point Tools* (PL. XL. FIG. 5.)

Are used for various purposes, as turning of mouldings, and the shoulders of screws, for which they are particularly useful ; they are sometimes employed in turning the flat ends of work.

----

§ 14. *Drills* (PL. XL. FIG. 6.)

Are used for making holes ; the work is fixed upon a chuck, but previous to this, the commencement of the hole is made with a point tool; the point of the drill is presented to this small cavity, and held in the line of the axis; then by pressing forward while the lathe is turning, the hole will be bored to any required depth; the drill should be drawn out once or several times, or the core will clog it, and prevent it from operating.

§ 15. *Inside Tools* (PL. XL. FIG. 7, 8, 9.)

Are employed for turning out hollows and cups of all descriptions, and have various forms, according to the curvature or angles of the work.

§ 16. *Screw Tools* (PL. XL. FIG. 10, 11.),

Are employed in cutting of screws of various sizes of threads. The work must first be turned truly cylindrical, then by applying the tool to the end, and pressing gradually with a uniform motion in the length of the axis, the screw will be produced

§ 17 *Flat Tools* (PL. XL. FIG. 14.)

Are used for turning cylindric or conic surfaces.

§ 18. *Square Tools.*

Are intended for brass turning only. In these, the cutting edges always terminate with right angles.

2 U

§ 19. *Triangular Tools*

Are used for turning iron and steel. They are of a triangular section, with three cutting edges, and are employed in turning planes or flat ends, also in the concave surface of the hollow bodies, as in cylindric and conic cavities.

§ 20. *Turning Gravers* (PL. XL. FIG. 13.)

Are used for turning steel and iron, in roughing out the work, though some works may be entirely finished by them. They are nearly the same shape as the tool used by engravers upon copper.

§ 21. *Parting Tools* (PL. XL. FIG. 14.)

Are used for making deep incisions, for cutting off a part of work, grooving, &c.

All these tools are bevelled or basilled from one side, except the chisel for soft wood, which is basilled from each side, and are all held upon a level with the axis, except the chisel.

§ 22. *Callipers*

Are used for taking the diameters of rotund bodies.

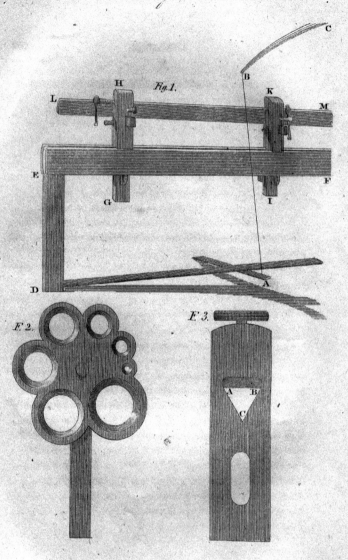

Turning.   _Plate XXXV._

Fig. 1.

F. 2.   F. 3.

W.S. Barnard. Sc.

§ 23. *Description of the Plates, with the Methods of Turning Elliptic Boards, Swash, and other Kinds of Work.*

PLATE XXXV.

The Pole Lathe.

Fig. 1 represents the pole lathe, as seen from the back.

A end of the foot-board or treadle.

A B the string to be coiled round the wood to be turned.

D E one of the legs, the other being hid in the view.

E F the shears or bed of the lathe formed of two pieces, with
a parallel space between.

G H, I K, the puppets, made moveable in the parallel space
and fixed below with wedges to any required distance, G H containing the fore centre, and I K that of the back centre. These
centres are tightened by means of screws.

L M the rest.

Fig. 2, large boring collar, with seven holes, from half an inch
to three inches and a half diameter.

Fig. 3, a boring collar for small work. The holes A B C may
be contracted at pleasure, by means of a sliding piece inserted in
a slip or groove parallel to the faces. The sliding piece is moved
by means of a thumb screw at D. The figure of the perforation
is an equilateral triangle, the lower part of the slider forming the
base of the said triangle ; then as a circle may be inscribed in an
equilateral triangle, the collar will fit all sizes of cylindrical bodies,
from the greatest size the perforation will contain, to the least, and
touch the body to be turned always in three points, which are all
that are necessary to steady the work in its revolution. This machine is generally constructed of iron.

PLATE XXXVI.

The Foot Lathe in its general Construction.

A B the treadle or foot-board.

a the manner of fixing the treadle to the floor.

C the crank hook, hooked into a staple, and the end of the piece A.

D the crank for turning the fly with the upper part of the crank hook formed into a collar for embracing the crank.

E the fly-wheel with several angular grooves cut in its circumference, in order to hold the band and keep it from sliding.

F the pillar for supporting the end of the mandrel.

G the puppet supporting the end of the mandrel, which holds the chuck.

H the right-hand puppet, containing the fore centre, which is tightened by means of a screw.

I, K the legs, the fly being supported by that of I, the other end is supported by an upright between the legs.

L the mandrel, showing the end of the spindle projecting over the puppet G, in order to receive the chuck.

M the rest, tightened below by means of a screw, and made so as to be fixed in any position to the chuck.

N a foot-board.

O several of the most useful tools employed in turning.

––––––––

§ 24. *Elliptic Turning.*

DEFINITION.

If there be a plane with any indefinite outline, and two inflexible right lines at right angles to each other, and if the plane be fixed to an axis at right angles therewith, and if the two inflexible lines be made to coincide with the plane, and be so moveable on

Turning. Plate XXXVI.

its surface, that one of them, which we shall call the primary line, may always pass through two fixed points in the plane, and through the point where the plane is intersected by the axis, and if the other transverse line be made to pass or slide along a given point, which is not attached to the plane, but would remain stationary even though the plane were in motion ; and if a secondary plane be fixed to the inflexible lines, parallel to the primary plane, then if the axis be carried round while the point in the transverse line is at rest, the primary plane will also be carried round, and every point in it will describe the circumference of a circle : the secondary plane will likewise be carried round, and will perform its revolutions in the same time as the primary plane and the axis, but being immoveably fixed to the rectangular lines, they will cause it to have both a progressive and retrogressive motion in the direction of the primary line in each revolution ; and lastly, if another point at rest be held to the surface of the secondary plane while in motion, it will either describe an ellipse, a circle, or a straight line. Hence the describing point will always be at the same distance from the centre or point, where the axis intersects the primary plane.

The eccentricity of the ellipse, or the difference of the axis, will be double the distance between the stationary point in the transverse line and the axis.

Instead of the stationary point, a circle may be placed with its centre in this point, and its plane perpendicular to the axis, and instead of the inflexible line moving to and fro along two fixed points in the plane, the diametrically opposite parts of the circumference may always touch a pair of parallel lines on the revolving plane.

PLATE XXXVII.

Illustrations. This Plate exhibits the various Positions of the Chuck for turning of Elliptical Work at every Eighth of a Revolution, according to the foregoing Definition.

Let A B and E F, No. 1, 2, 3, 4, 5, 6, 7, 8, be the two inflexible lines, intersecting each other in I, at right angles, and let C, D be the two fixed points. Let A B be denominated the primary line, and E F the secondary line, and let the lines A B and E F at right angles, taken as a wnole, be called a transverse ; also let C represent a primary point, and let the describing point be taken at G, in the line drawn through C D produced : now in all positions of the chuck, the primary line A B is always upon the point C, and E F upon D ; having premised this in general, suppose, before the machine begins to start, that E F, No. 1, the secondary line coincides with E G, and the point G with *o, o* being in the plane of the figure to be described ; then because A B always passes through C, the points I and C will be coincident, A B being then at right angles to E F.　Let us now suppose the motion to commence, and let it perform an eighth part of a revolution, as at No. 2, the describing point G still remaining in the same position with respect to C and D, viz. in the right line to C D G, then the point *o* will now be at a distance from the point G, and a part G *o* of the curve will be described by the fixed point G, also the point i will be above the line C D G : now let the motion proceed, and describe another eighth as at No. 3 ; then the point *o* being always in the line E F produced, E F will be at a right angle with the fixed line C D G, and A B coincident with C D G, and the point which was last at G, will now be at I.　In like manner, when another eighth has been performed, as at No. 4, the point *o* has performed three-eighths of a revolution, the point 1 is in a line drawn from the point C, perpendicular to the fixed line C D G, and the point 2, which was at G, in No. 3, is situated between 1 and G.

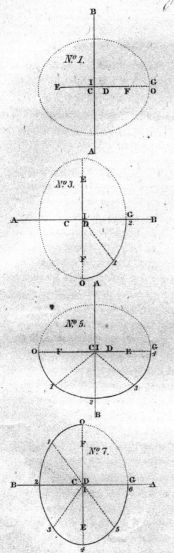

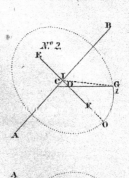

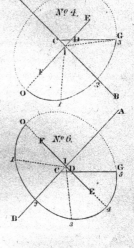

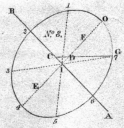

In this manner, by continuing the motion, the whole curve will be generated. No. 5 shows the curve, when half a revolution has been described, No. 6, five-eighths; No. 7, six-eighths, or three quarters; and No. 8, seven-eighths.

Here it may be proper to observe, that the angles performed by the revolution of the machine, are very different from the corresponding angles, formed by lines drawn from the centre of the ellipse, to the describing point, and to the extremity of the curve at its commencement.

From what has been said, it is easy to conceive that the operation of elliptic turning is nothing more than that of the ellipsegraph or common trammel, with this difference, that in the operation of turning, the ellipse is described by moving the plane, and keeping the point steady, but in forming the curve by the ellipsegraph, the plane of description is kept steady while the point is in motion. The transverse A B E F is the same as the grooves in the trammel cross, and the line C D G the trammel rod : here the cross and plane of description move round together, but fixed to each other, and the trammel rod C D G is held still or immoveably confined : in the trammel, the board and cross are fixed together, and held while the trammel rod C D G moves with the points C and D in the grooves.

To set this machine therefore, it is only to make C D equal to the difference of the axis.

PLATE XXXVIII.

Shows the relation between the foregoing diagrams and the chuck. Let K L M N be the face of a board representing the plane, which is fixed to the axis of the machine. And let O P Q R be another board made to slide in the board K L M N, each two points O and K, L and P, M and Q, N and R, coinciding at this moment: K L M N will therefore represent a wide groove in the board; as this groove may be of any width, we may conceive the breadth to be very small or nothing, and may therefore be represented by a groove or by the line A B parallel to K N and L M, and in the middle of the distance between them. Instead of supposing the point D always moving to and fro in the line E F, we may suppose a circle, or the end of a large cylindric pin moving in a very wide groove T U V W across the slider O P Q R. Now therefore, all the differences between these diagrams and those in the former plate, are only wide grooves in place of lines passing longitudinally through the middle: for the line A B is always conceived to move reciprocally from the one side to the other of the board K L M N: now it is the same thing whether one straight line slide longitudinally upon another fixed line, or whether a bar of any breadth move in a groove of the same breadth, or whether a straight line in reciprocal motion always pass through two fixed points.

No. 1 shows the chuck, as in the first diagram of the last plate: No. 2 as No. 2, No. 3 as No. 3, and No. 4 as No. 4 of the said plate. Any farther explanation is conceived as unnecessary. It now remains to explain how the chuck is connected with the machine, and how the parts are connected with each other.

The end of the spindle of the mandrel passes through a stout upright, and projects over it with a convex or male screw, to which is fixed the board K L M N with the faces at right angles to the axis: a circular ring or end of a very large pin is attached to the said side of the upright, so that the ring or pin may be

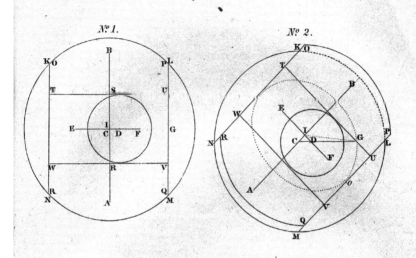

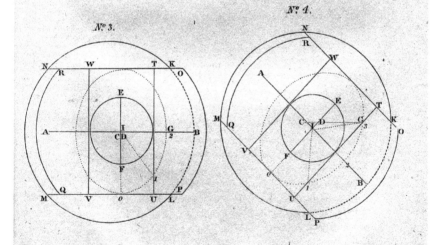

fixed at any required distance from the axis of the spindle, and that its axis and the axis of the mandrel may always be in the same horizontal line or plane.

The wide groove K L M N is made on the inside of the board next to the face of the upright, and equal in breadth to the diameter of the cylindric pin, and the slider may either move in a groove upon one side or the other, or move in mortises, but in whatever mode the reciprocal motion of the slider is performed, the groove in the slider must always be made from the inside, so that the board which is fixed to the axis must be cut away for that purpose, in order that it may fit upon the ring or pin, and since the work to be turned is fixed upon the outside of the slider, the slider must be flush both outside and inside, or the slider may project on the outside.

It has been mentioned, that it is of no consequence what the boundary line of the board is, neither does it signify what the combination of the parts are that form the chuck, so that the same principle of motion is performed. The parts exhibited in this plate, show the most simple form of the principle, and therefore the diagrams are better calculated to afford instruction. In some chucks, the principle is almost concealed by a complication of parts, which, though not necessary in forming the motion, are essential in the practice : for this reason, by continual working, if the parts were only of the most simple forms, when the grooves and pins wear, the truth of the motion would be destroyed without any remedy to rectify it. In the best constructed chucks, the board which is screwed upon the end of the mandrel is a frame, which is variously constructed by different people, but the parts of it which form the sides of the grooves, may be brought nearer together by means of screws, and thus the sliders and the cylindric ring or pin may move exactly in the grooves.

The drawing of the chuck, and the manner in which it is connected with the machine, is exhibited in Plate V. to the explanation of which we must refer our reader for further information ; the geometrical principle, and the manner in which it is combined with, and their relation to the parts in practice, being all that is intended to be explained in this place; and indeed this is almost the whole that can be done. The practice can never be obtained from any written description, but only from the actual exercise of the art itself, so that any farther attempt, besides the uses of the tools, which we have already given, would be needless : one thing only is to be observed, that in turning several ellipses, the circumferences will be nearly parallel, as the difference in their several axes is the same.

PLATE XXXIX.

Fig. 1 is a view of the end of the machine; the principal parts shown in this view are

A the pulley of the mandrel.

B and C sides of the frame supporting the pulley.

D frame for the rest to slide in.

E and F legs supporting the frame D.

G and H continuation of B and C below the frame of the rest.

I nut and screw under the frame of the rest.

K the elliptic chuck with two grooves, through which the knobs of the slider pass, and are connected on the outside by a strong bar of iron, which is screwed upon their ends. This also shows the screw for fastening the board to which the work is fixed This frame is strongly braced to the roof, in order to keep it steady.

P the rest.

Q the piece by which the rest is fastened.

Fig. 2 a view of the inside of the chuck, containing the parts N and O : this side of the chuck being placed against the side C of the frame, Fig. 1.

N the board containing the slider O, showing the end of the

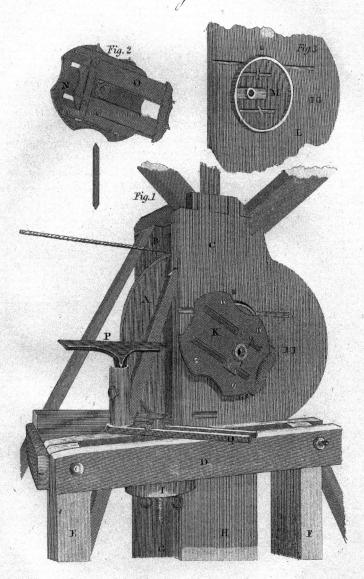

Fig. 2

Fig. 3

Fig. 1

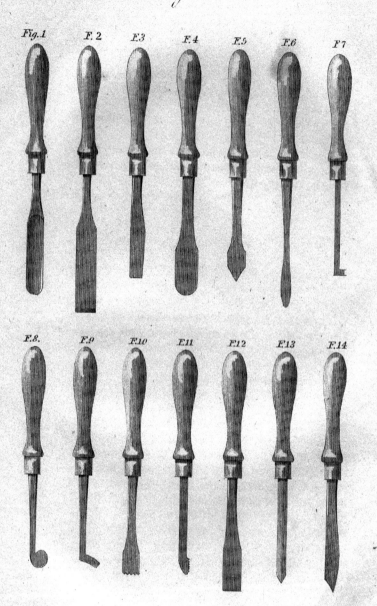

Turning   Plate XL.

Fig.1   F.2   F.3   F.4   F.5   F.6   F.7

F.8.   F.9   F.10   F.11   F.12   F.13   F.14

screw which is fixed in the mandrel; the board N revolves round a centre, while the slider O not only moves round, but has a longitudinal motion to and fro in the part N.

Fig. 3 a view of the outside of the mandrel frame, showing the parts L and M.

L a part of the side C of the mandrel frame, showing the ring M which is fastened to it and which causes the reciprocal motion of the slider O in Fig 2

———

PLATE XL.

TOOLS.

Fig. 1 the gouge for roughing and traversing the work.

Fig. 2 the chisel used in smoothing cylindric, conic, and convex surfaces after the gouge.

Fig. 3 right-side tool.

Fig. 4 round tool.

Fig. 5 point.

Fig. 6 drill.

Fig. 7 inside tool for angular work, all the sides being made to cut occasionally as well as the upper side of the hooked part.

Fig. 8 inside tool for concave curved work.

Fig. 9 inside tool for turning a solid sphere within a hollow one.

Fig. 10 screw tool for the convex or male part.

Fig. 11 screw tool for the concave or female part.

Fig. 12 flat tool.

Fig. 13 turning graver.

Fig. 14 parting tool.

For the particular properties and uses of these tools, see articles where they are particularly described.

§ 26.   *To turn a Hollow Sphere.*

First turn the convex surface, on which draw two great circles
at right angles to each other ; then the line joining the intersection
of these circles, is an axis of the sphere, which will divide each
circle into two equal parts or into half circles : divide each semicircle into two equal parts, and each circle will be divided into
quadrants.   Upon each of the intersections or poles, with a centre
bit, bore a cylindric hole, with its axis tending to the centre of
the sphere, to such a depth as to leave the solid space between
the two bores equal to the diameter of the cylindrical bores, or
something less, with the same centre bit upon the division of each
semicircle ; bore holes tending to the centre as at first, and of the
same depth : there will be now six holes, then if the axis of any
two be fixed in a straight line with that of the mandrel, with the
convex surface of the sphere in a hollow chuck, then the interior
surface may be turned out to a certain extent, and formed by
means of the instrument shown in Plate XL. Fig. 8 : take the
sphere out of the chuck, and place the hollow part thus turned in
the chuck, fixing it fast therein with the axis in the same straight
line with that of the mandrel, then turn the opposite hole in like
manner.   Proceed in like manner with each two remaining pairs
of opposite holes : in turning, the hollows must be so large as to
penetrate each other, and leave only so much of the solid to connect the sphere with the core, as is sufficient to support the latter :
then each of the eight connecting parts must be sawn through
close to the core, and as the core is less than either of the holes,
it may be taken out, and the connecting pieces may be sawn off
with a bent saw close to the concave surface, and thus you will
have the hollow sphere required.

§ 27. *To turn one Sphere within another.*

Find the centres of the cylindrical holes as before, then bore each of the holes to an equal depth, so that its axis may tend to the centre of the sphere, and that the thickness between each pair of opposite holes may be equal to, or something more than the diameter of the required interior sphere ; then fixing the axis of each hole in the axis of the mandrel, with the tool represented in Plate XL. Fig. 8. turn a part of the interior surface of the outer sphere, and a part of the convex surface of the interior sphere, and thus leave eight connecting parts, which are each to be cut with a bent saw, close to the convex surface of the interior sphere, and to the concave surface of the exterior sphere.

If the cylindrical holes are perforated or bored quite through, a series of spheres may be turned within each other by the same means, but the diameter of the least must be greater than that at the bore ; it would be best to begin the operation with the most interior sphere, and after this the next, and thus in succession till the one next the exterior one be loosened. In performing the cylindrical excavations, the diameter of each hole may be continually less, and in proportion to the diameter of each of the internal spheres.

In the same manner may a cube be turned within a sphere, instead of turning the surface of the interior solid spherical, it is only turning it flat by means of an inside tool, which has its cutting edge straight, and at a right angle with it.

———

§ 28. *Conclusion.*

Many kinds of turning may be performed by making the axis of the work to be turned to slide progressively, or with a reciprocal motion through two collars, as given points according to a certain law, while the body continues to revolve uniformly. If the axis proceed with a uniform motion, and a tool be pressed to the surface, the tool will cut a spiral line on the said surface.

If a single crank be fixed to the end of the mandrel, and the end of the crank made to touch an inclined plane while the body is in motion, the point of a sharp tool being pressed upon the surface, and kept stationary by means of the rest, a line will be cut or described on the surface of the wood, and this line will be the circumference or perimeter of an ellipse, which will have the proportion of its axes in the ratio of radius to the sine of the plane's inclination. If the surface of the body to be turned be straight, and the cutting edge of the tool be always held equi-distant from the axis, the body itself will be turned into a cylinder, and all its sections perpendicular to the axis will consequently be circles.

If the surface of the body be turned into mouldings, the work is denominated *swash work,* which was much in request in former times, for bodies standing upon the rake, or upon an inclined plane, as in the balusters of staircases, but is now entirely laid aside.

An indefinite variety of subjects or figures may be obtained by turning, by different regulations of the mandrel, by making the crank slide upon various surfaces, or by other methods of regulating the axis in a direction of its length.

INDEX

AND

EXPLANATION OF TERMS

USED IN

TURNING.

N. B. *This Mark* § *refers to the preceding Sections, according to the Number.*

A.

Axis, an imaginary line passing longitudinally through the middle of the body to be turned, from one point to the other of the two cones, by which the work is suspended, or between the back centre and the centre of the collar of the puppet, which supports the end of the mandrel at the chuck.

B.

Back Board, that part of the lathe which is sustained by the four legs, and which sustains the pillars that support the puppet bar. The back board is only used in the best constructed lathes. In the common lathes, the shears or bed are in place of the back board, § 5.

Back Centre. *See Centres,* and § 5

Band, § 5. *See also Cat-gut.*

Bearer, that part of the lathe which supports the puppets. § 5.

Bed of the Lathe, the same as bearer, which see.

Boring Collar is the machine having a plate with conical holes of different diameters; the plate is moveable upon a centre, which is equidistant from the centres or axes of the conic holes; the axes are placed in the circumference of a circle. The use of the boring collar is to support the end of a long body that is to be turned hollow, and which would otherwise be too long to be supported by a chuck. *See Plate* XXXV. *Fig.* 2.

C.

Callipers, compasses with each of the legs bent into the form of a curve, so that when shut, the points are united, and the curves being equal and opposite, enclose a space. The use of the callipers is to try the work in the act of turning, in order to ascertain the diameter or diameters of the various parts. As the points stand nearer together at the greatest required diameter than the parts of the legs above, the callipers are well adapted to the use intended.

Cat-Gut, the string which connects the fly and the mandrel, § 5.

Centres, are the two cones with their axis horizontally posited for sustaining the body while it is turned, § 5.

Cheeks, the shears or bed of the lathe as made with two pieces for conducting the puppets.

Chisel, a flat tool, skewed in a small degree at the end, and bevelled from each side, so as to make the cutting edge in the middle of its thickness, § 9.

Chuck, a piece of wood or metal fixed on the end of the mandrel for keeping fast the body to be turned, § 6.

Circular Turning, § 2.

Collar, a ring inserted in the puppet for holding the end of the mandrel next the chuck, in order to make the spindle run freely and exactly, § 5.

Collar Plate. *See Boring Collar.*

Connecting Rod. *See Crank Hook.*

Conical Points, the cones fixed in the pillars for supporting the

body to be turned; that on the right hand is called the fore centre, and that on the left hand, the back centre, § 5.

CRANK HOOK, sometimes called also the connecting rod, as it con
nects the treadle and the fly, § 5.

CRANK, the part of the axle of the fly, which is bent into three
knees or right angles, and three projecting parts; one of the
parts is parallel to the axis, and has the upper part of the crank
hook collared round it, § 5.

D.

DRILL, § 14.

E.

ELLIPTIC TURNING, § 25.

F.

FEET, the horizontal pieces on the floor which support the legs of
the lathe, § 5,

FLAT TOOLS, § 17.

FLY WHEEL, § 5.

FOOT LATHE, § 5.

FOOT WHEEL, or FLY, the wheel or reservoir for preserving and
continuing the motion when the force applied by the foot is not
acting, § 5.

FORE CENTRE, that on the right hand. *See Centres*, § 5.

G.

GOUGE, the tool for roughing out the work, § 8.

I.

INSIDE TOOLS, § 15.

L.

LATHE, the machine for holding and giving motion to the body to be turned, when the requisite force is applied.

LATHES in general use, § 3.

LEFT-SIDE TOOLS, § 11.

LEGS, the uprights mortised into the feet for sustaining the upper part of the lathe, § 4, 5.

M.

MANDREL, that part of the lathe which revolves the body when turned in a chuck : the pole lathe has no mandrel, § 5.

MANDREL FRAME, the two puppets which hold the mandrel ; a hardened steel collar being fastened in the fore puppet, and a screw with a conical point in the back puppet.

N.

NOSE, that part of the spindle of the mandrel which projects over the puppet to receive the chuck, § 5.

O.

OVAL CHUCK, § 25.

P.

PARTING TOOLS, § 21.

PIKES, now called conical points, which see.

PILLARS, the uprights fixed at the ends of the back board, for sup porting the bed of the lathe or puppet bar, § 5.

PITCHED, is the placing of the work truly upon the centres.

POINT TOOL, § 13.

POLE, an elastic rod fixed to the ceiling of the turner's shop for re-acting by means of the string upon the treadle against the pressure of the foot; the foot draws the string downwards, and the pole exerts its force in drawing it upwards, and consequently should have no more elasticity than what is sufficient for this purpose, as the overplus would only tire the workman, § 4.

POLE LATHE, § 4.

PULLEY, § 5.

PUPPET BAR. *See Bearer.*

PUPPETS, the upright parts for supporting the mandrel, the one on the right being called the fore puppet, and that on the left the back puppet; the screw is fixed on the one, and the mandrel collar on the other puppet, § 5.

R.

REST, the part of the lathe which sustains the tool while turning, § 4, 5.

RIGHT-SIDE TOOLS, § 10.

ROUGHING OUT, is the reducing of the substance by means of the gouge, to prepare the surface of the body for smoothing.

ROUND TOOLS, § 12.

S.

SCREW, the conical points or centres as made with a screw, in order to tighten the work; the screw or screws ought to be kept so tight, that there should be no play, otherwise the work may be in danger of flying out, § 5.

SCREW TOOLS, § 16.

SHEERS. *See Cheeks or Bed of the Lathe.*

SLIDER, § 25.

SQUARE TOOLS, § 18.

STRING, that which connects the treadle and the pole in the pole lathe, and in the foot lathe it passes round the fly-wheel and the pulley of the mandrel in order to turn the latter.

SWASH WORK, § 29.

T.

Tools, § 7.

Traversing, is moving the gouge to and fro in roughing out the work.

'Treadle, the part of the lathe by which the foot communicates its force, and gives motion to all the other moveable parts, § 5.

Triangular Tools, § 19.

Turning in General, § 1.

Turning Gravers, § 20.

W

Wabble is the shaking of the work in the act of turning, because it is not fixed truly upon the centres.

There are several other terms which are common to smithing and turning. See the Index and Explanation of the Terms to those articles.

THE

STEAM-ENGINE,

ITS USES, &c.

Description of the Nature of Steam, of the Principle of the Steam-Engine, of its various Modes of Construction, and of its several Parts, showing how they act upon each other.

So much has been said by a host of able writers in admiration of the power and utility of STEAM, that it would be little better than a waste of words to expatiate upon what is already self-evident without the aid of eloquence. But it is necessary to those who desire fully to appreciate and understand the nature of this elastic fluid, to make themselves acquainted with the laws by which its operations are governed; the laws appertaining to its generation and condensation; its increase and decrease of elasticity, resulting from and depending upon an increase or decrease of temperature. Without this indispensable knowledge, it will be impossible to comprehend the value of the many ingenious and important changes, not only in form, but in principle, which the power of steam has undergone, since its first rude introduction as an agent of mechanical force. The present object, therefore, is to simplify and illustrate this part of the subject as much as possible.

The two component elements of steam are water and heat, or caloric. Steam has been defined to be the invisible vapour of water which is given off by it at all known temperatures under certain degrees of pressure; a definition which, though correct, is not in accordance with the popular idea of steam. It has also been defined as the offspring arising from the union of the two elements, fire and water; and some conception may be easily formed of its stupendous properties, when it is stated that its expansive force is found by experience to be much greater than that of gunpowder. *Vapour* is nearly synonymous with *steam*, the latter term, however, being usually limited to express the vapour from water, and the former being more general in its application. Aqueous vapour, in its perfect state, is transparent and colourless, consequently invisible; and it is only when partially mingled with the air, or having touched substances cooler than itself, that it becomes vesicular, and consequently visible. The moist, white vapour, therefore, composed of an infinite number of vesicles, or small globules, is not, as generally supposed, perfect steam, but steam which has been deprived of a portion of caloric.

Water.

The expansive quality of water, when subjected to the action of fire, has been variously estimated. It is said that one cubic inch of water will expand into about a cubic foot of steam, occupying a space 1800 times greater than it does in its original state of water. But Mr. Davies Gilbert made the estimate considerably less, and after repeated experiments, was satisfied that it increased its original bulk when converted into steam only 1330 times;

whereas, by Dalton's experiments, it was computed to occupy 1711 times the bulk it did in its original state.

Water itself is a compound substance, consisting of two gases, oxygen and hydrogen, so called from two Greek words, the former signifying to generate acids, the latter to generate water. It owes its fluidity to the latent heat, or caloric, which it contains, and the absence of which is the cause of its being reduced to the state of ice.

The weight of water is of importance. One pint will weigh one pound; a cubic foot of water, will, therefore, weigh 1000 ounces, or 6½ lbs. avoirdupois, being about 48 lbs. for a cylindrical foot. It is 816 times heavier than atmospheric air.

It has been ascertained that the time required to convert a given quantity of water into steam, is six times greater than that required to raise it from the freezing to the boiling point. Fluid, exposed in an open vessel to the action of fire, cannot, however great the heat applied, be made to indicate a higher temperature than that at the boiling point. Steam will be evolved in greater or less quantities, according to the degree of heat applied, but the temperature will continue the same as that of the water; a phenomenon first investigated by Dr. Black.

Heat.

Heat is a quality, or principle, the nature of which is not known, but which is inherent in all substances. Chemically, it is called *caloric*, in order to distinguish heat itself as a matter, from its effects: by this term it is known as the cause of heat, that is, as distinct from the glow, or warmth, imparted by one body hotter than others, adjoining to it, until an equilibrium is established, and to this equilibrium it has a constant tendency. The properties of heat, if fully investigated, though constituting a highly interesting subject, would occupy too great a space for the present work. But it must be stated, that the degree of heat contained in specific bodies has been calculated by various tables: that in water, is generally computed at 60 degrees of Fahrenheit as unity.

The *capacity for heat* and the *specific heat* of bodies have been sometimes confounded; for this reason, that as one body absorbs more heat than another, in order to raise its temperature a certain number of degrees, it is said to have a greater capacity for heat.

The distinctive terms may, however, be better defined thus: that *capacity for heat* should be limited to the relation between the whole quantity of heat in bodies of the same temperature; and *specific heat* to express the quantity necessary to produce a certain change of temperature. The increase, or abstraction, of heat produces a change frequently in the state of bodies, as, for example, subtract heat from water and it becomes ice; infuse heat into ice and it becomes fluid; and, by raising the temperature, water becomes steam or vapour. Steam has a greater capacity for heat than water, and water greater than ice.

The heat or caloric existing in all bodies consists of two portions, materially differing from each other. The one is denominated *sensible heat*, or that which is in a free or disengaged state, and which is perceptible to the sense; the other, *latent heat*, or that which is fixed in any body, and not evident to our sensations, or to the thermometer. Other terms have been given to this description of caloric, by Professor Pictet, who calls it *caloric of fluidity*, and *caloric of vaporization*. Caloric in its latent state is inactive, or dormant; released from this it affects the sense of feeling, and the thermometer, as though it had never been latent.

The doctrine of latent heat was first discovered or illustrated by the celebrated Dr. Black; and it has been erroneously asserted that Watt was indebted to this discovery for the improvements which he made in the steam-engine; but this assertion has been abundantly proved to be without foundation, by a long letter addressed by Watt himself, towards the close of his life, to Sir David Brewster, who was then engaged in editing a new edition of the works of Dr. Robison, by whom the error was originally propagated. Mr. Watt says, "I have always felt and acknowledged my obligations to him (Dr. Black) for the information I had received from his conversation, and particularly for the knowledge of the doctrine of latent heat," but he continues, "I never did, nor *could*, consider my improvements as originating in those communications."

Numerous experiments have been made by eminent men for the purpose of ascertaining the different degrees of latent heat existing in aqueous vapour, or steam; and the following results will show the difficulty of arriving at the desired point.

Latent Heat of Aqueous Vapour.

	Fahrenheit.
Black	800°
Ure	888°
Southern	945°
Watt	950°
Clement	990°
Lavoisier	1000°
Thompson	1016°
Rumford	1021°

Dr. Ure's experiments were undertaken in 1817, and published in the *Transactions of the Royal Society* for 1818. The following table furnished by him, but the numbers of which were afterwards corrected as under, by Mr. Tredgold, gives the mean result for water, and some other fluids.

Liquid.	Specific gravity.	Temperature of the water in the basin.			Boiling point.	Heat of conversion into vapour.
		Beginning.	End.	Difference.		
Water	1·000	42·5°	49°	6·5	212°	942°
Alcohol	0·825	42	45	3	175	425·50
Sulphuric Ether	0·7	42	44	2	112	302·60
Oil of Turpentine	0·888	42	43·5	1·5	316	146·0
Petroleum	0·75	42·5	44	1·5	306	150·0
Nitric Acid	1·494	42	45·5	3·5	165	517·0
Ammonia......	0·978	42	47·5	5·5	140	840·0
Vinegar	1·007	42·5	48·5	6		870·0

Elastic Power of Steam, Atmospheric Pressure, &c.

When the thermometer of Fahrenheit indicates a temperature of 212°, the force of steam exactly equals the pressure of the atmosphere; but the annexed table, given by Woolfe, exhibits the results produced by higher degrees of temperature; although it has been since discovered, by experiments, that the figures contained in the last column are erroneous; an observation, however, that does not apply to the preceding columns, which are the material ones for practical guidance.

Steam predominating over the pressure of the atmosphere upon a safety valve, if its elastic force be equal to		pounds per square inch, requires to be maintained by a temperature equal to		degrees of Fahrenheit thermometer; and at these respective degrees of heat, steam can expand itself to		times its volume, and yet continue equal in elasticity to the pressure of the atmosphere.
	⎧ 5 ⎫		⎧ 227½ ⎫		⎧ 5 ⎫	
	⎪ 6 ⎪		⎪ 230¼ ⎪		⎪ 6 ⎪	
	⎪ 7 ⎪		⎪ 232¾ ⎪		⎪ 7 ⎪	
	⎪ 8 ⎪		⎪ 235¼ ⎪		⎪ 8 ⎪	
	⎪ 9 ⎪		⎪ 237½ ⎪		⎪ 9 ⎪	
	⎨ 10 ⎬		⎨ 239½ ⎬		⎨ 10 ⎬	
	⎪ 15 ⎪		⎪ 250¼ ⎪		⎪ 15 ⎪	
	⎪ 20 ⎪		⎪ 259½ ⎪		⎪ 20 ⎪	
	⎪ 25 ⎪		⎪ 267 ⎪		⎪ 25 ⎪	
	⎪ 30 ⎪		⎪ 273 ⎪		⎪ 30 ⎪	
	⎪ 35 ⎪		⎪ 278 ⎪		⎪ 35 ⎪	
	⎩ 40 ⎭		⎩ 282 ⎭		⎩ 40 ⎭	

The importance attached by scientific men to this subject, inasmuch as it was extremely desirable to obtain the most accurate knowledge, based on satisfactory principles, of the elastic power of steam at higher temperatures, influenced the French government to solicit the members of the Institute, at Paris, to engage in a new series of experiments relating to the object in view. The inquiry was undertaken by MM. Arago, De Prouy, Du Long, and Girard, who in 1830 made their Report to the Institute, by which eminent body it was presented to the Minister of the Interior. It was, of course, a document abounding in materials of great value; but we can only advert to it in general terms in this work; and those who desire, for practical purposes, to become better acquainted with its contents, must refer to the Report itself, which may be easily procured in this country.

The elastic power of steam, when not in contact with its water of generation, but is detached from it, has an increase of temperature which does not increase its density, but merely its elastic form, the density remaining the same.

Velocity of Steam.

It is difficult to arrive at any fixed rule of computation in determining the degree of velocity with which steam will rush into a vacuum, or a fluid more rarified than itself: as so much would depend upon circumstances, as to the hindrances it might have to encounter through pipes, or other apertures, in making its escape. For this reason (the little practical good that could be attained by such investigations) they have not been encouraged; experience has been found the best guide. In general terms, it has been reckoned that the passages in low pressure-engines should be in diameter as $\frac{1}{5}$ to the diameter of the piston, and consequently the area about $\frac{1}{25}$; or say about an inch per horse for the area of the passages.

Condensation of Steam.

Without the means of condensing this subtle vapour with ease and rapidity, the steam-engine, in comparison with what it is now, would be impracticable and powerless for any valuable purposes. Its inefficiency for so long a period is mainly ascribable to the want of this knowledge. It did not enter into the minds of the Marquis of Worcester, Moreland, or indeed any of the early machinists, who turned their attention to improvements on the first rude models.

The process of the condensation of steam is fortunately simple, and will be readily comprehended. When steam comes into contact with a body colder than itself, they have a natural tendency to affinity with each other: that is, they both become of an equal temperature, the one imparting cold and the other heat to its neighbour; or, in other words, the heat finds its equilibrium. Thus, the colder a body is, or its temperature being the same, the greater its surface, the more rapidly the temperature of the hot body will be reduced, and, in the case of steam, the more rapid its condensation. Two things should, therefore, be observed in a steam-engine,—1st. That in the process of condensation, the surface of the body through which the condensation is to proceed, should be made as great, and its temperature as low, as possible. 2d. The condensing body should be a good conductor of heat, for by the possession of this quality the process of condensation is the more rapidly effected. Water is the best conductor of heat yet discovered, and is consequently used for the purpose of condensation in steam-engines. Who first discovered the effect of water here mentioned is involved in doubt, Savery having claimed that distinction, and Desaguliers having conferred the honour on Newcomen.

Water, however, though universally used as the best known medium for the purpose of condensation of steam, is attended by some inconveniences well understood by practical men. As all the water once used for injection must be removed from the condenser, this cannot be done without the water bringing with it a portion of atmospheric air, which must also be removed, an operation requiring the assistance of an air-pump, of considerable force, which is a reduction of so much useful effect. Among the innumerable experiments constantly being pursued by clever and ingenious mechanics, many have endeavoured to remedy these defects, with little, if any success; that is, to produce condensation, without using the injecting water. Patents have been taken out for the same purpose, but seem to have shortly been abandoned by the patentees; and if any new and successful method had been discovered, it would doubtless soon come into general adoption. It cannot, notwithstanding, be concealed, that many excellent inventions fail for want of due notice and encouragement, or the want of means on the part of the owners.

Mechanical Portions of the Steam-Engine, with their Relative Bearings to each other.

Three things should be required in a steam-engine; first, that it should occupy the smallest possible space; secondly, that it should produce the greatest power for its size; and thirdly, that it should consume the least quantity of fuel.

Steam-engines are described under three general names, as denoting so many different principles or modes of construction and working: viz.—

1st, The High Pressure Steam-Engine.

2d, The Mean, or Low Pressure ditto.

3d, The Expanding ditto.

To which should be added, the high pressure condensing engine, working expansively, a combination of these, long in use in Cornwall, and lately introduced by Mr. Wicksteed at the East London Water-works, where it is working with great success.

Having already given in our former pages a brief historical account of all the different fire or steam-engines, recorded to have been invented, from the earliest period of which there is any mention made of a machine worked by steam; and subsequently having explained the nature and power of steam itself, the next object is to describe the mechanical parts of the engine, by means of which such almost magical effects have been produced; and we know not that we can do this better than by proceeding, step by step, with a description, in chronological order, of the various improvements made in it, from its first simple invention to the present day.

The contrivance, or apparatus of Hero, invented about 2000 years ago, indicated the first idea of which we have any knowledge of the application of steam for the purpose of producing motion. The following wood-cut, with the subjoined description, will convey a sufficient notion of this crude conception.

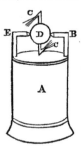

A, represents a caldron in which the steam is generated by means of a fire kindled underneath, but concealed; B, the hollow support by which the elastic vapour passes to the only apertures through which it can escape, marked CC. The ball, or globe, marked D, is connected with the hollow tube, or pipe, by a steam joint, but still such as to admit of freedom of action; and the opposite side of the globe with the solid arm, marked E, so as to allow of the free rotation of the ball. As the steam rushes out from the apertures above described, and encounters the atmosphere, the reaction produces a rapid rotation.

Brancas's invention was even yet more simple. It consisted of a vessel representing the head of a negro, filled with water. It was supplied with a small tube proceeding from the mouth, and intended to give motion to a-wheel placed opposite to it, the elastic vapour issuing from which set the wheel in a continual whirl. But the invention will be better understood by the following cut.

A, represents the vessel containing the water; B, the furnace; C, the tube through the aperture of which the steam issues; and E, the wheel which the action of the elastic vapour, coming in contact with it, sets in motion.

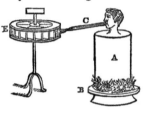

The Marquis of Worcester's improvement on the steam-engine is the next entitled to notice; but, from the meagre particulars he has left of his invention, there is little to be added to the account given by himself.

It has been a question whether the Marquis of Worcester had discovered the property of steam to condense, or not. Different opinions have been held upon this subject, and it does not appear to be as yet finally settled, as the writings of the noble author throw little or no light upon it. Some persons have thought that he made the discovery about the middle of the seventeenth century; but others have maintained, with more probability, that he was acquainted only with the expanding power of steam, and not with its property of rapid condensation, upon which so much of the value of the steam-engine depends. Nothing, it has been observed, can be more obscure and unintelligible than the account which this mysterious writer has left of his invention; although there can be no doubt that he obtained some extraordinary results by the operation of the power of steam.

Mr. Tredgold, in his *Treatise* on the *Steam-Engine*, denies that the Marquis of Worcester had any knowledge of the contracting power of steam, and gives the plan of an engine such as that which he supposed to be indicated by the noble author's description; but as this idea is merely conjectural, it is unnecessary in a work intended for popular use to enter farther upon the inquiry.   One thing is certain, that the Marquis of Worcester succeeded in reviving attention to the capabilities of steam; he prevented the matter from sinking into oblivion; he stirred up the minds of men of learning and genius to its farther consideration; and ultimately led the way to the most gigantic results.

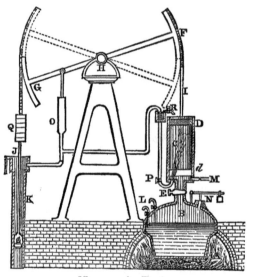

Newcomen's Engine.

Newcomen's engine may be considered as the foundation of all the improvements since made in the steam-engine, and which smoothed the path for his successors.

The principle and parts of this engine are simple, and may be readily understood by the following description, and reference to the engraving.

A, represents the furnace; B, a section of the boiler; C, of the cylinder in which the piston D moves, air-tight, up and down between the points D and d; E, the steam-cock to regulate the supply of steam from the boiler to the cylinder.   The beam F, G, moves on the centre or axis H.   At the end F, a chain is secured to the upper extremity of the arch-head, which passes round it to the lower extremity, where it is fastened to the piston rod I, completing the connection between the cylinder and the beam; the other end of the beam, G, is in like manner connected to the pump-rod J, working in the pump-barrel, K.   O represents a small pump working with a packed piston, which supplies water to be injected into the cylinder at the cock, P.   Q, weights on the pump-rod, to drive it down into the pump, and to give a preponderance to that end of the beam.   L, represents the gauge-cocks, to determine the level of the water in the boiler; M, the escape-pipe through which the injected water and condensed steam pass off from the cylinder. N, is the safety-valve, opening outwards, to regulate the pressure of the steam in the boiler; it is adjusted by moving the weight along the lever, to or from its axis, on the principle of the common steelyard weighing-machine.   R, a jet of water which may be let on to the top of the piston, to keep it air-tight.

To set the engine to work, the fire having been lighted, and the boiler filled to the proper level, it will be necessary to close the steam-cock, E, until the steam in the boiler has attained sufficient power to raise the safety-valve, then turn on the steam-cock, when the steam will rush into the cylinder, but will be immediately condensed by the cold surfaces opposed to it, and which will continue to be the case until the cylinder has acquired the same temperature as the steam.   The air, the cylinder contained before the steam was let in, and the injection water, together with that produced by the condensation of the steam, will be driven out through the pipe M.   The cylinder now being full of steam, turn off the steam-

2 s

cock, and turn on the injection-cock, the cold water spouting into the cylinder will immediately condense the steam, and produce a vacuum, when the piston will be driven to the bottom of the cylinder, by the pressure of the atmosphere, and consequently the first stroke of the pump effected. The steam-cock being again turned on, the steam, together with the excess of weight on the pump-rod, will raise the piston to the top of the cylinder, and by continuing the process just described, the engine will be kept in action.

Professor Brande, however, in his lectures at the London Institution, introduced so simple and ingenious an instrument for the purpose of illustrating the operations of an atmospheric steam-engine, that a representation of it will materially assist the reader in comprehending the foregoing description of Newcomen's engine.

The glass tube and ball is shown with its piston A; the rod being hollowed, and closed by a screw B. If steam be generated by the spirit lamp C, the air will speedily be expelled; and after this is effected, the screw B may be closed, and a working stroke is obtained by artificial condensation.

If the reader has perused the above description attentively, he will be prepared to investigate the still more extensive, important, and efficient improvements made in the steam-engine by the superior mechanical genius of Watt, and which, with a few additions, will bring us down to the present day.

Hitherto our observations have applied to atmospheric steam-engines only; but Watt introduced an entirely new principle. It has been seen that in Newcomen's engine, it was the weight of the atmosphere acting upon the piston that pressed it downwards. Watt proposed to make the steam itself depress the piston, instead of by atmospheric pressure, an object in which he completely succeeded.

By this method, the substitution of steam for atmospheric pressure, he obtained a most important advantage, as he considerably augmented, or rather doubled, the power of the engine; because, in an atmospheric steam-engine, the force acting upon the piston must depend solely upon the weight of the atmosphere above it; but if the elasticity of the steam be increased to twice the amount of the atmospheric pressure, an engine, upon the latter principle, will raise a column of water of twice the weight that one upon the former principle could.

Watt's second great improvement was that of employing a separate vessel for condensing the steam, by which he produced several considerable advantages, more particularly a saving of nearly half the fuel; because, by surrounding the cylinder with a case, it is always preserved at the temperature of boiling water, or at nearly the same as that of the steam; so that it loses nothing in the process of condensation. .

The character of the steam-engine was thus completely changed; its power vastly augmented, and its force more easily regulated. For more than fifty years, Newcomen's or the atmospheric engine maintained an undisputed supremacy, and even long afterwards it was preferred by many persons to the more modern one; but the superiority of Watt's engine rapidly superseded the use of all others. Although, however, it is to the practical genius of that eminent man that the world is indebted for the principal improvements in the steam-engine, it is probable that those advantages would have been lost to mankind but for the penetration, public spirit, and noble munificence of his partner, Mr. Boulton.

Baron Dupin, in reference to this topic, says, "Watt's engine was, when invented by him, but an ingenious speculation, when Boulton, with as much courage as foresight, dedicated his whole fortune to its success." "Men," continues Dupin, "who devote themselves to the improvement of industry will feel in all their force the services that Boulton has rendered to the arts and mechanical sciences, by freeing the genius of Watt from a crowd of extraneous difficulties which would have consumed those days that were far better dedicated to the improvement of the useful arts."

The above, however, were not the only improvements made in the steam-engine by Watt, as he also introduced the governor, parallel motion, &c., by which it was placed more completely under control, and thus rendered the more perfect. But as we shall give a minute description of the steam-engine now in general use, accompanied by a cut, it is unnecessary to give any further account of his inventions in this place.

The double acting steam-engine is that of which we purpose giving a full and detailed notice, because if the principle and parts of that be thoroughly comprehended, there will be little difficulty in understanding others. The frontispiece represents an engine upon the most approved modern construction, and the following letter-press description will explain all its various parts and bearings.

***

The Boiler.

The Boiler is undoubtedly the most fundamental part of the steam-engine, not only on account of the nature of the functions it has to discharge, but also by reason of the various apparatus attached to it, by means of which its temperature is regulated, and thereby the liability to accidents diminished. The most fatal accidents may, and often do, result from imperfections in the boiler itself, or from the neglect, or mismanagement, of the persons whose duty it is to attend to it. If more steam be generated, at a given rate of speed, than is requisite for that speed, the boiler being overcharged will burst; unless some provision be made to guard against such a consequence. To obviate, as far as possible, any danger from this cause, several ingenious appendages are attached to the boiler, the first of which is the *safety-valve,* which allows the escape of any superfluity of the vapour. But another precaution is necessary to prevent the unnecessary consumption of fuel and waste of power; and this suggested the idea of the *self-acting damper,* which is so adjusted to the engine, that when the action of the steam becomes too great, the damper descends and checks the heat of the furnace; and, when too small, the damper, by rising, increases the draught and intensity of action. To some engines are also attached internal safety-valves for additional security against any sudden and unusual condensation of steam in the boiler. Next to the damper is the self-acting *feed-pipe,* the use of which is also to regulate the heat in the boiler, to prevent the metal being burnt and destroyed by the immoderate action of the fire, and also to avoid laying a foundation for many serious disasters. *Gauge-cocks* form another attachment to the boiler: they have "pipes descending into them to such depths, that one of them opens just above the proper surface-level of the water, and the other just below it, so that when the water is in proper quantity and the first of these cocks is turned, it will yield steam, and the other being turned, it will yield water; but if the water is too high in the boiler, both pipes will be immersed in, and give out water, and if too low, they will both give steam; the engineer, therefore, by this means, has it always in his power to ascertain the state of his boiler, and whether or not the self-acting damper is properly performing its duty."

It may be proper to observe here that more recent improvements have taken place in steam-engine boilers; among which may be mentioned the introduction of cylindrical boilers and serpentine tubes—by which four great objects were obtained,—increased safety, considerable space, a great saving of fuel, and an accession of power. H. M. steamer *Echo* was fitted up in this manner in 1831, and the machinery was found to answer excellently.

The dimensions of boilers is a question that can be interesting to few but practical readers, and moreover it could not be adequately considered within our prescribed limits; for which reasons it is better altogether to omit the subject.

———

Safety - Valves.

The object of the safety-valve is, by permitting, as occasion may require, the escape of the steam, to promote the safety of the engine and the security of human life. It is made in a variety of forms, according to the opinions, or experience, of the makers. It will be sufficient here to notice only one kind, namely, that which is called the steel-yard safety-valve. The construction is as follows:—a support projects from the side of a short tube, to which a lever is attached, carrying a weight at its other end, and to the lever is fixed the valve, which opens when the pressure of the steam from within exceeds that to which the weight is adjusted, or rather to the part of the lever to which the weight is appended. But as the valve will sometimes stick, the most frightful consequences may, in that case, be apprehended; and, in fact, extensive loss of life has been incurred from this circumstance; insomuch, that a parliamentary inquiry was some years since instituted into the subject. Various expedients have been tried to prevent the sticking of the valve; and it is always best to have a box or cover over it to prevent its derangement by the ignorance and meddling of strangers. The valves are to the steam-engine what the lungs are to the human body.

———

Pistons.

The power of the steam-engine, in a great measure, depends upon the working qualities of the piston, which is that part of the machinery which is attached to the rod called the piston-rod, and which ascends and descends in the cylinder by the action of the steam. As it is necessary that the cylinder should be kept steam-tight, it is the province of the piston to perform that operation; at the same time, it is important that as little friction should be produced between the piston and the side of the cylinder as possible. In Newcomen's, and other engines, this was effected by means of water being kept floating on the upper surface of the pistons; and this was one of Watt's first difficulties. He at length determined, after many experiments with different kinds of packings, and different unguents for diminishing the friction, that hemp, or tow, lubricated with tallow, best answered the purpose. Notwithstanding, however, the utmost pains taken with this mode of packing, the drawback upon the power of the engine, from friction, is considerable, being generally estimated at 288 pounds per foot on the contact surface of the piston. Much ingenuity has, therefore, been exercised in order to obviate so serious a deficiency.

Hemp-packed Pistons.

The hemp, or common packed piston, was for many years universally employed as a matter of necessity, or as the best that had been devised; it is even still, in some instances preferred; and occasionally it becomes quite necessary to have recourse to it, upon temporary emergencies, particularly in steam-vessels at sea, even where metallic pistons are used, in case of the latter getting out of repair, or from being damaged. But the packing of the piston always causes an inconvenient delay of several hours. The following is the mode in which it is done. The bottom of the piston is first fitted as accurately as possible to the cylinder, leaving it, however, freedom enough to rise and fall through the whole length;

the part of the cylinder immediately above this is from one to two inches, according to the size of the engine, less all round than the cylinder, to leave a circular or annular space, into which unspun long hemp, or soft prepared rope, is wound as evenly and compactly as possible, to form the packing. A plate or cover is then put over the top, having a projecting ring to fit over the lower part, to complete the upper side of the space for packing, being compressed upon it by means of several screws. Both the upper and lower part of the space round the piston to contain the packing is a little curved, that the pressure produced by the screws on the packing may force it against the inside surface of the cylinder, into as close contact as possible. The operation of re-packing, however, if it were attended with no other disadvantage, is a sufficient objection to it as an uniform or general practice.

Metallic Expanding Piston.

The first projector of a metallic piston was the Rev. Mr. Cartwright, about fifty or sixty years since; but he was unable to bring it into successful use, and the old mode of packing continued to prevail. Various attempts were made to improve Cartwright's piston, but most of them were failures, or at least not entitled to much encouragement, until Mr. John Barton, about 1815, took out a patent for a metallic expanding piston, which has since been introduced into the government service, and very extensively adopted in other public and private establishments. Mr. Barton has made several improvements in his piston since it was first projected, by which its utility has been greatly increased. Without encumbering our pages by describing it in its original form, we shall confine ourselves to a description of it in its present more perfect state.

Barton's metallic piston is represented by the annexed plan and section. It is composed of a solid cylindrical cast-iron body A, having a conical hole B, to receive the enlarged end of the piston rod, to which it is secured by a bolt. A space or groove is formed round the body of the piston, to receive metallic segments marked E, which are spread asunder by four triangular wedges G, of the same metal as the segments, acted on by eight spiral springs of tempered steel. These springs are inserted in cylindrical cavities at both ends, in order to render them secure from bending, and yet allow them to play freely. With the same view, each spring has a piece of steel within it, a little shorter than the spring. In pistons made for high-pressure steam, there are three grooves formed round the exterior part of the segments; the middle designed to hold oil or grease to lubricate the rubbing surfaces. The upper and lower grooves are for hoops of tempered steel, having a forked loose joint in each; these hoops are nicely fitted to the grooves, and when the piston is in its place their jointed ends meet.

Barton's Lubricators.

This is a very simple but exceedingly useful invention. By the old mode of oiling the bearings of steam-engines, much oil was unnecessarily consumed and spilt about, causing a filthy appearance. To rectify this, Mr. Barton invented his patent Lubricator. It consists of a vessel either cylindrical, urn-shaped, quadrangular, or in any other form required. A tube passes through the body of the vessel, elongated three or four inches, or less, according to the size, below the bottom, and by which it is fixed to the bearing.

Through this tube two or three threads of worsted are passed, one end overhanging the top of the tube in the vessel, and the other hanging out from the bottom of the tube. The vessel being filled with oil, the oil is absorbed by the worsted, and by capillary attraction conveyed to the bearing to which the lubricator is attached. The supply is regulated by means of a screw and valve at the top; the screw being made tight, the valve presses upon the top of the tube in the vessel and stops the supply; but by relaxing the pressure of the screw, the oil continues to flow in any proportion that may be desired. Thus the Lubricators render a constant and well-regulated supply of oil to all those parts of the engine where it is required, save the engine-men much personal risk, and a great deal of labour, which is especially the case when the engine is working at great speed.

Cylinder and Condenser Gauge.

In order to ascertain the state of the engine at any time, two barometer gauges are employed, one showing the elasticity of the steam in the boiler, and the other, the rarefaction of the vapour in the condenser. The nature of these instruments is too well known to need any further description of their uses.

Parallel Motion.

This is a beautiful contrivance, for which we are indebted to the celebrated Watt. The object is to connect the end of the beam to the piston rod in such a manner, that whilst the end of the beam rises and falls in the arc of a circle, the piston shall move up and down in an exact straight line. It is evident, if the end of the piston rod was to be fixed to the end of the beam by a pivot, that it must follow the annular course traversed by the beam head, which would rock it to and fro till it either broke the rod or destroyed the packing in the box of the cylinder lid through which it passes.

With the more varied application of steam, it became also necessary that the communication should be so arranged that the piston should either push up the beam, pull it down, or both alternately, as in the double-acting engine.

The apparatus he adopted for effecting these purposes is represented on the arm A of the beam in the frontispiece. The beam moving on its axis C, every part of the arm moves in the arc of a circle of which C is the centre. Let B be the point which divides the arm A C into equal parts, A B, and B C; and let D E be a straight rod equal in length to C B, and playing on the fixed centre D. The end E of this rod is connected by a link E B with the point B. If the beam be supposed to move up and down on its axis C, the point B will move in a circular arc, of which C is the centre, and at the same time the point E will move in an equal arc, of which D will be the centre; accordingly the middle point F will move up and down in a straight line.

Then if a link A G, equal to B E, be connected at A to the end of the beam by a pivot at B, and its other end attached to the rod G E, and the piston rod T, by the same pivot G, the point G with the piston rod will move up and down in a straight line, as was shown of the point F, to which it is similar, although it travels over twice the space. Thus, we have two points moving in straight lines F G; to F is attached the air pump rod, and to G the piston rod.

The Governor

Is a throttle valve placed in the tube which conveys the steam from the boiler to the cylinder, an ingenious contrivance of Watt to regulate the supply of steam. For a time it was left to the engine-man to adjust, but he afterwards arranged a plan by which the engine was made to regulate it, and with more certainty and accuracy than could be attained by the old system. He effected this by communicating the motion of the fly-wheel by the strap K to a vertical shaft, to which two balls were suspended by rods, marked H in the frontispiece; as the shaft turned round, carrying the balls with it, they by the centrifugal force spread out from their centre, more and more, as the speed increased. Connected with the balls was a lever, which gradually closed the valve h as the balls rose, shutting off part of the steam, and consequently diminishing the power. If, on the contrary, the speed of the fly wheel was diminished, the balls fell toward the axis, and the opposite effect ensuing, the supply of steam was increased and the speed restored.

The Double-acting Engine.

When Watt had, by his great ingenuity and perseverance, improved upon the atmospheric engine of Newcomen, and rendered it suitable for all purposes of raising water, he conceived the idea of applying the force of steam to manufactures generally. To accomplish this, it became necessary that the beam should be driven by the steam in its ascent, as well as in its descent, and he first proposed to effect it by employing two cylinders, one placed under each end of the beam, to work alternately, so that one piston would be descending by the full power of the steam exactly when the power was suspended in the other, consequently a universal force would be exerted on the beam.

He, however, soon laid this aside for another arrangement still more simple, with which he produced the same result with one cylinder, by alternately admitting the steam above and below the piston, the steam at the same time being drawn off by the condenser from the other side. Thus the steam forces up the piston, and forces it down alternately, and is therefore called the Double-acting Steam-engine, an illustration of which will be seen in the frontispiece of this work.

Having given some account of the principal portions of the steam-engine separately, we will now endeavour, by describing them collectively, to give the reader an idea of the relation and influence they bear to one another. The steam from the boiler passes to the cylinder through the pipe S, and the quantity admitted in a given time is regulated by the valve h, called the throttle valve, which is acted upon, or is under the direct influence of the governor.

Attached to one side of the cylinder are two square, hollow vessels or boxes marked U V, having communication with the cylinder by a hole from the middle part of each, called the port holes; these boxes are supplied with four valves, the upper one in each being the induction valve, for the admission of steam, and the lower one in each the exhaustion valve, by which the steam passes off from the cylinder to the condenser. These valves move in pairs, that is, the upper induction and the lower exhaustion move together, and the upper exhaustion and the lower induction also at the same time. All these valves are worked with one lever marked W. Under the cylinder is placed the condensing apparatus, consisting of two cylinders O and N, immersed in a cistern of cold water, the former of which, the condenser O, has a pipe called the injection-pipe entering into it, with a head like the rose of a watering-pot, while the other end is open to the cistern, thus establishing a communication between the interior of the condenser and the cold water in which it is immersed. This pipe is supplied with a cock to regulate the passage of the water through it. The other cylinder, N, is called the air-pump, and is furnished with a close-packed piston, having a valve in it opening upwards, and operating like the bucket of an ordinary pump. This air-pump piston has an upright rod, with its lower end secured to the piston, and the upper end connected to the beam at the point F. The bottom of these vessels, O and N, are connected by a passage having a valve in it opening towards the latter.

The principal cylinder, marked Q in the frontispiece, is constructed similarly to that employed for the single-acting engine, and therefore does not require any further description.

We will now suppose the piston to be at the top of the cylinder, and the whole area of the cylinder below the piston to be filled with steam. To set the engine going we must turn on the injection-cock, which will throw a jet of cold water into the condenser; then by raising the lever W, the lower exhaustion valve will be opened, and the steam that occupied the cylinder will pass off to the condenser, and be immediately reduced to water, causing a vacuum below the piston; at the same time the upper induction valve will be opened, and the steam being admitted on the upper surface of the piston, will force it down into the vacuum, thereby producing an effectual downward stroke of the beam. To produce an upward stroke, we have only to depress the lever W, when the valves that were open will be closed and the other pair opened. The steam escaping from above the piston to the condenser, a vacuum is again formed into which the piston is driven by the force of the steam as it is admitted beneath it by the lower induction valve.

When the steam is reduced to water in the condenser, a certain quantity of air becomes disengaged. This air, and the water produced by condensation, together with the water that was thrown in by the injection pipe, must be removed, to perform which is the office of the air-pump, and is effected by every upward stroke of that end of the beam to which the pump rod is attached.

It should be observed in the Double-acting Engine, condensation is constantly going on, therefore the injection-pipe must always be flowing, but requires to be regulated according to the quantity of steam supplied to the cylinder in a given time; so that only the exact quantity of cold water necessary to reduce the steam to water at the temperature of 212 degrees may be admitted.

Of the hot water removed by the air-pump, a quantity is carried back to the boiler by the pipe L, sufficient to keep it properly supplied.

Trusting the foregoing treatise has been sufficiently clear and comprehensive to give the reader a general idea of the admirable contrivances by which the elastic force of steam is rendered such an active and powerful agent, we will proceed to show how the force is communicated to machinery when a rotary motion is required.

At the opposite end of the beam to that at which the piston-rod is exerting its power, is attached a rod marked X, by the pivot Z. The lower end of this rod works a crank or shaft on a longitudinal beam that is capable of revolving. Now it is obvious that this rod can only exert power in the direction of its length, consequently can only push and pull the crank; but as it is necessary that the crank should go round its centre, an agent must be employed to assist the rod when its force is least effective. This agent is the fly-wheel. The fly-wheel is a heavy disc or hoop, balanced on its centre, and so connected with the machinery, that it turns rapidly with it, so as to receive its motion from the impelling power. The momentum which it acquires during the descent of the connecting rod, it will retain, by virtue of its inertia during the time that the power of the rod is least effective, and will carry it past the perpendicular position, and bring it again within the influence of the rod. Thus the horizontal shaft will continue to revolve, impelled by a powerful and uniform force, and which may, by the ingenuity of the mechanic, be varied and adjusted to any of the numerous uses to which the mighty, irresistible power of steam is applied.

The High-pressure Engine.

The principle of engines so called depends on the power of steam to expand itself very considerably beyond its original bulk, by the addition of a given portion of caloric, thus acquiring a considerable elastic force, which, in this case, is employed to give motion to a piston. One of the greatest advantages attendant on employing the repellant force of steam, as in this form of the engine, consists in an evident saving of the water usually employed in condensation; and this, in locomotives for propelling carriages, is an object of considerable importance. The invention is ascribed to Papin; but it afterwards received great improvements from the genius of Watt. But the first persons who employed the high-pressure engine to advantage, were Messrs. Trevithick and Vivian, as they found it admirably adapted for the purpose of propelling carriages. In this case the steam, after having performed its office, was thrown off into the air; and the condenser, together with the necessary supply of cold water which must have accompanied it, was by this means dispensed with. For the purpose of motion, the high-pressure engine certainly possesses considerable advantages, not the least of which are cheapness and portability.

TABLE OF CONTENTS.

PRACTICAL GEOMETRY.

Practical problems performed on the ground.

CARPENTRY.

JOINERY.

CONTENTS.

353

BRICKLAYING.

MASONRY.

SLATING.

PLASTERING.

PAINTING IN OIL.

SMITHING.

——

TURNING

THE STEAM-ENGINE.

TABLE

Showing the Pages opposite which the Plates are to be placed.